PREPARATION GUIDE FOR

LABORATORY INVESTIGATIONS

FOR BIOLOGY

PREPARATION GUIDE FOR

LABORATORY INVESTIGATIONS

FOR BIOLOGY

Jean Dickey

Clemson University

The Benjamin/Cummings Publishing Company, Inc.
Redwood City, California • Menlo Park, California • Reading, Massachusetts
New York • Don Mills, Ontario • Wokingham, U.K. • Amsterdam • Bonn
Sydney • Singapore • Tokyo • Madrid • San Juan

- *Sponsoring Editor:* Don O'Neal
- *Assistant Editor:* Kim Johnson
- *Production Coordinator:* Teresa Thomas
- *Copy Editor:* Judith Hibbard
- *Composition:* Richard Kharibian
- *Artist:* Karl Miyajima
- *Cover Designer:* Yvo Riezebos

Many of the designations used by manufacturers and sellers to distinguish their products are claimed as trademarks. Where those designations appear in this book, and Benjamin/Cummings was aware of a trademark claim, the designations have been printed in initial caps. See page 114 for a list of trademark acknowledgments.

ISBN 0-8053-0933-0

1 2 3 4 5 6 7 8 9 10—VG—98 97 96 95 94

The Benjamin/Cummings Publishing Company, Inc.
390 Bridge Parkway
Redwood City, California 94065

Preface

Lab preparation is an essential ingredient of successful laboratory learning. *Preparation Guide for Laboratory Investigations in Biology* contains all the information needed to set up the laboratory for the lab topics in the Lab Investigations manual. The preparation instructions for the Alternative Exercises are included in the packet with the exercises.

Each Prep Guide chapter contains a list of materials needed for each exercise, including a complete description of any item that is not self-explanatory. Catalog numbers are given to supplement the description of many items. In most cases it is not necessary to order the item from any particular company; comparable materials are available from numerous sources. A list of vendor addresses appears at the end of the Prep Guide on page 107. A Recipes section describing how to prepare all solutions is included in each lab topic that requires one.

Students will work in teams of 2–4 students for most lab exercises. The course instructor will tell you how many teams to prepare for per lab section. In many exercises there are materials that can be shared among teams. Work with the course instructor to plan how to arrange the materials in your lab rooms for the most efficient use of resources.

Investigative Labs

Several of the lab topics are "investigative." After performing an introductory experiment, students design and perform their own investigations based on the technique they have just learned. Preparation for investigative labs includes a basic kit for each team and additional materials they may request. Complete instructions are given for assembling the basic kits. Suggestions for additional materials are given as explicitly as possible. Be sure to communicate with the person in charge of the course regarding materials that should be available.

Planning

A Planning Guide at the end of this book (page 109) will help you anticipate your needs for the entire course, organize your ordering, and plan your preparations. This handy checklist contains each piece of equipment needed as well as all specimens, solutions, and chemicals, listed by their corresponding lab topic (including the Alternative Exercises, which appear in gray shading). Some materials, especially pieces of equipment such as Spec 20s, are optional; the Instructor Notes in the Lab Investigations manual give details. Materials that can be found in any laboratory, such as test tubes, beakers, pipets, and so forth, are not listed in the Planning Guide.

There is also a Summary of Materials at the end of each lab topic to provide a quick overview of the items required for that particular lab. These summaries do not include the additional materials students may request for investigative labs.

Safety

All staff members involved in lab preparation should be aware of OSHA regulations. You should be trained in general laboratory safety procedures, including handling, storage, and disposal of hazardous substances and biohazards. A few reminders for safety in the teaching labs follow.

- Containers of solutions that are potentially hazardous for students should be clearly labeled when supplied to the lab rooms.

- Provide waste containers for substances that should not be poured down the drain.
- Provide containers for broken glassware.
- Lab rooms should be equipped with eyewash stations, safety shower, first aid kit, fire extinguisher, and kits for mopping up spilled acid, base, and mercury (if you use mercury thermometers).
- If there is a phone in the lab, post emergency numbers. If there is no phone, post the location of the nearest phone along with emergency numbers.

Animal Use

Most institutions now have policies regarding the use of living vertebrates in any type of laboratory. A protocol must be filed and approved before the animals are purchased, so allow time for this procedure when preparing for Lab Topics 13 (Animal Diversity II) and 19 (Animal Behavior). If you are not familiar with the animal use regulations at your institution, ask the course instructor or departmental administrator.

Questions/Feedback

Preparing biology laboratories is a complex task. I have tried to compile a preparation guide that makes your job easier. Please call, write, or e-mail if you have questions, need additional information, or find errors that should be corrected in future editions. I'd also like to hear which features of the Prep Guide you find useful and suggestions for additional information to include.

Jean Dickey
Biology Program
330 Long Hall, Clemson University
Box 341902
Clemson, SC 29634-1902
(803) 656-3827
Internet: dickeyj@clemson.edu

CONTENTS

1 The Process of Scientific Inquiry 1

2 Tools for Scientific Inquiry 3

3 pH and Buffers 6

4 Macromolecules 14

5 Using the Microscope 21

6 Enzymes 25

7 Cellular Respiration 29

8 Photosynthesis 38

9 Mendelian Genetics 42

10 Microorganisms and Disease 47

11 Plant Diversity 54

12 Animal Diversity I 64

13 Animal Diversity II 69

14 Digestion 74

15 Circulation 81

16 The Sensory System 83

17 Plant Structure 88

18 Flowers, Fruits, and Seeds 94

19 Animal Behavior 100

20 Foraging Strategy 104

Vendor Addresses 107

Planning Guide 109

Trademark Acknowledgments 114

The Process of Scientific Inquiry

Exercise 1.1: The Black Box

For this exercise, each pair or team of students will have a small container with one or two objects sealed inside, a plastic bag that contains most (but not all) of the objects that have been placed in the containers, and an empty container (not sealed).

Small matchboxes or plastic film canisters can be used as the containers. Seal with tape (black electrician's tape works well) once the objects are inside.

The objects used can be any small, common items, such as a paper clip (small or large size), thumbtack, screw, penny, and so on.

At least one container should have a small feather in it, but do not put a feather in the plastic bag.

One container should have a safety pin in it (the same size as a regular paper clip) but do not put a safety pin in the plastic bag.

The plastic bags should contain all the items that are sealed into the containers *except* the feather and safety pin. The contents of all the plastic bags should be the same.

The instructor should have one single-edged razor blade for each pair or lab team. The razor blades should not be available to the students until the instructor is ready for them to open their containers.

Suggested combinations in containers:

Container	Items
1	regular paper clip, thread
2	safety pin
3	penny, feather
4	thumbtack, thread
5	feather, safety pin
6	penny, large paper clip
7	thumbtack, safety pin
8	regular paper clip, feather
9	bobby pin, thumbtack
10	penny, thumbtack
11	large paper clip
12	bobby pin, safety pin

The plastic bag can contain: regular paper clip, large paper clip (use metal paper clips, not plastic), penny, thread, thumbtack, and bobby pin. That is, the plastic bag should have all the items that are in *all* the containers *except* the feather and safety pin.

It is not necessary for instructors to know what is in each container, but if they want to, you can number the containers and provide the instructor with a key.

The sealed containers will be opened by the students during the lab session. If your labs are scheduled consecutively, you need to plan to have a new set of sealed containers ready to go while you recycle the opened set.

Exercises 1.2, 1.3, and 1.4

No materials are required for these exercises.

Optional

An instructor who wants to have students do a step test experiment in this lab topic may request additional materials. See Lab Topic 15 (Circulation) in this guide for information on devices to measure blood pressure.

Summary of Materials

	Per team of students
Labware	
plastic bag	1
film canisters or matchboxes	2
miscellaneous items	2 of each
single-edged razor blade	1

Tools for Scientific Inquiry

Exercise 2.1: Scientific Notation

No materials are needed for this exercise.

Exercise 2.2: Making Measurements

Activity A: Measuring Length

Materials for each pair of students

- Ruler: Either 6" or 12". Rulers must have markings for metric units along one side and English units on the other side.

Activity B: Measuring Mass

Materials for each pair of students

- Balance: Consult with the instructor about which balances to use for this lab topic. They should be the same type of balance that students will use for the rest of the course. Students can take turns, so it is not necessary for each pair of students to have a balance to themselves—but the more balances that are available the better.
- Weigh boat.
- Set of objects to weigh: Consult with the instructor about the set of objects for students to weigh. They should span the range of the balance's capability. Examples:

Item	Weight (in grams)
dropper bulb	1.45
glass Petri dish(medium)	16.06
rubber stopper (#9)	44.11
nut and bolt	53.37
full box of slides	363.09
500 mL water in plastic bottle	578.88

It is convenient to put each set of objects in a small box.

Activity C: Measuring Volume

Materials for each pair of students

- 5-mL blowout pipet (TC): Also known as serological pipets (for example, Fisher #13-675K; Carolina Biological Supply #73-6512).
- 5-mL delivery pipet (TD): Also known as Mohr measuring pipets (for example, Fisher #13-665K; Carolina Biological Supply #73-6276).

❋ NOTE: The instructor may want you to use only one kind of pipet or the other throughout the course.

- Pi-pump for 5-mL pipet (green): The pi-pumps referred to here and throughout the lab manual are plastic plunger-type pipet filling devices (Carolina Biological Supply #73-6870 and 73-6875). If your lab is already equipped with some other type of pipet filling device, the instructor may prefer that you substitute those.
- 250-mL flask containing water tinted with food coloring: Students will use the water to practice measuring with pipets and grad cylinders. Each flask should have at least 100 mL in it.
- 250-mL beaker labeled "waste container": Students will measure colored water into this container and then pour it back into the flask.
- 100-mL graduated cylinder
- Optional: 1-mL pipet: Ask instructor.
- Optional: pi-pump for 1-mL pipet (if 1-mL pipets are used)

Activity D: Measuring Temperature

No materials are required for this exercise.

Exercise 2.3: Data Presentation

It is useful to have graph paper available in the labs for students to use, but the instructor may ask them to provide their own. If you supply it, it may be cheaper to photocopy sheets of graph paper than to buy them. Paper that is ruled 10 squares per inch is a good type to use, but check with the instructor to see if a specific type is preferred. Each student would use about two to four sheets.

Exercise 2.4: Interpreting Information on a Graph

No materials are required for this exercise.

Summary of Materials

	Per team of students
Disposables	
graph paper (optional)	2–4 sheets
Labware	
250-mL beaker	1
250-mL flask	1
100-mL graduated cylinder	1
1-mL pipet (optional)	1
5-mL blowout pipet	1
5-mL delivery pipet	1
pi-pump for 1-mL pipet (optional)	1
pi-pump for 5-mL pipet	1
15-cm (6") or 30-cm (12") ruler	1
weigh boat	1
set of objects to weigh	1
Equipment	
balance	1 (teams can share)

LAB TOPIC 3

pH and Buffers

In this lab topic, students learn a method for investigating pH (Exercise 3.2) and a method for investigating buffering capacity (Exercise 3.3). In Exercise 3.4, students design an experiment using one of these methods. In Exercise 3.5 they perform their own experiment. Some instructors may do Exercises 3.1, 3.2, 3.3, and 3.4 one week and Exercise 3.5 the following week.

Students will work in teams throughout the exercises. The instructor will specify the number of students per team (usually 2–4) and will tell you how many teams to prepare for per lab section. The quantities of materials needed by each team are the same regardless of team size.

Exercise 3.1: Introduction to Acids, Bases, and pH

Materials for the instructor

- 2 50-mL beakers
- Bottle of phenol red indicator (~20 mL will be used in each lab section; see Recipes)
- Small drink straw (1 for each lab section, as it will be discarded after use)

Exercise 3.2: Using Red Cabbage Indicator to Measure pH

Activity A: Making a Set of Standards
Materials for the instructor (only one set needed for a lab section)

- ¼ red cabbage
- Knife
- Cutting board
- 2 1-L beakers
- Hot plate
- 10" × 10" piece of cheesecloth, 4 layers thick (1 for each lab section, as it will be discarded after use)
- Small funnel
- 12 dropping bottles labeled "cabbage juice"

Materials for each student team

- Test tube rack
- 15 test tubes: Seven test tubes are used to make a set of standards that will be used throughout the lab period. If you need to conserve glassware, give each student a few additional test tubes to perform the tests in Activity B; these can be rinsed and reused.
- Wax marker
- Buffers, pH 2, 4, 6, 7, 8, 10, and 12: ~5 mL each buffer per team. See Recipes section. A set of

buffers can be shared by more than one team of students so that you don't have to prepare 7 bottles for each team. Put caution labels on bottles containing pH 2 and pH 12 buffers.

- 8 5-mL pipets: A separate pipet is used to measure each buffer. If teams are sharing sets of buffers, they can also share pipets. Tape a test tube to the side of each bottle to use as a pipet holder. Label the pipets to avoid contamination. The additional pipet is needed by each team to measure their cabbage indicator.
- Pi-pump for 5-mL pipet
- 20 Parafilm squares (used to cover test tubes). Cut the regular size (2" × 2") Parafilm squares into quarters.

Activity B: Comparing pH of Beverages and Stomach Medicines
Materials for each student team

Supply the following in dropping bottles. Each team will use ~1 mL.

- 7UP (Sprite or another colorless soda can be substituted)
- White wine (do not use a wine cooler)
- White grape juice
- Seltzer water (club soda): Do not use the flavored varieties.
- Sodium bicarbonate: 0.5% solution
- Milk of Magnesia: Dilute approximately 1:1 with tap water to make solution easier to pull up into medicine dropper.
- Maalox: Do not use the flavored (tinted) varieties; solution must be colorless. Any other colorless antacid can be substituted. Dilute approximately 1:1 with tap water to make solution easier to pull up into medicine dropper.

Exercise 3.3: Using the pH Meter to Determine Buffering Capacity

Materials for each student team

- pH meter
- Wash bottle of distilled water
- 250-mL beaker: Fill with water or buffer to soak electrode when not in use.
- 600-mL beaker
- Magnetic stir plate (optional): This is not an absolute necessity, but the experiments can be performed more rapidly and conveniently using one (e.g., Fisher #11-501-4S).
- Small stir bar (only if stir plate is used): Be sure the sink drains have screens in them to catch the tiny stir bars when students discard their solutions.
- Glass stir rod (if stir plate not used)
- Bottle of 0.1N NaOH (~20 mL, 1 per group): Put a caution label on this bottle.
- Bottle of 0.1N HCl (~20 mL, 1 per group): Put a caution label on this bottle.
- 2 1-mL pipets
- Pi-pump for 1-mL pipets
- 100-mL graduated cylinder
- 100-mL beaker
- Bottle of water: It makes no difference to the experiment whether tap or distilled water is used, but label the bottle to tell which it is.
- Bottle of buffering solution used in Exercise 3.2 (each team will test one solution and will use ~80 ml)

Exercise 3.4: Designing an Experiment

No materials are needed for this exercise.

Exercise 3.5: Performing the Experiment and Interpreting the Results

Each student team will design an experiment using materials from either Exercise 3.2 or Exercise 3.3. They will be investigating the pH values of various substances. Consult with the instructor regarding the number of teams to prep for. Students who are using the red cabbage technique (Exercise 3.2) will need extra test tubes and pasteur pipets with rubber bulbs. Students may also request additional materials. Consult with the instructor regarding the schedule and the manner in which these items will be requested and the types of items the instructor would like to have on hand. The Instructor's Annotated Edition of the lab manual suggests that students could bring their own samples for their investigations. Some possibilities include:

Exercise 3.2

Each team would use solutions from one or two of the following categories, and would use approximately 2 mL of each solution that they test. For this method to work, all samples must be clear or relatively colorless (a light yellow tint is acceptable).

- Beverages: Examples include white grape juice, apple juice, grapefruit juice (but not ruby red), Gatorade and other sports beverages, Sprite, 7UP, milk, buttermilk, white wine, wine coolers, beer of different brands.
- Shampoos: Examples include any shampoos and/or conditioners that are white or clear. Light yellow color (for example, baby shampoo) is okay, but avoid green.
- Hygiene products: Examples include different brands of toothpaste, deodorants, or mouthwash.
- Cleaning solutions: Examples include soft soaps, different brands of laundry or dishwashing detergents, cleaning solutions such as Pine-Sol and Lysol, drain openers (dilute 1:1), and scale removers such as Lime-A-Way. Be sure to put caution labels on caustic substances such as drain openers and Lime-A-Way.
- Water samples from different sources: Examples include distilled water, deionized water, bottled water, tap water (from lab, dorm, or home), drinking fountain water, water from nearby pond, lake, ocean, stream, or river.
- Pain relievers: Examples include aspirin and aspirin-substitutes (e.g., Tylenol, Motrin), different formulations of aspirin (e.g., buffered, with caffeine). Students should make a solution by crushing 1 tablet in 100 mL water.
- Antacids: Examples include any antacids except Maalox and Milk of Magnesia (used in Exercise 3.2). If thick liquids are used, dilute approximately 1:1 with tap water so solutions can easily be pulled up in a medicine dropper. Students should make a solution from tablets (for example, Tums or Rolaids) by crushing 1 tablet in 100 mL water.

Exercise 3.3

- 0.1N HCl and 0.1N NaOH: Each team will need approximately 20 mL of 0.1N HCl and 0.1N NaOH in addition to the amount used in Exercise 3.3. Each team that does this type of investigation will use approximately 80 mL of each of the substances they test, and a typical investigation will compare two substances. Since pH meters are used in this method, the color of the solution doesn't matter. However, be sure students don't use solutions that would damage the electrodes. We have learned the hard way that deodorants are especially destructive.

- Beverages: any of the examples listed for Exercise 3.2 might be used.
- Stomach or pain medicines: In addition to the examples listed for Exercise 3.3, students might use Alka-Seltzer or Pepto-Bismol.

Recipes

pH 2 Buffer

Each team needs ~5 mL and an additional 80 mL per lab section.

To make 1 L:

Dissolve 7.46 g KCl (potassium chloride) in 200 mL distilled water. Dilute up to 250 mL with distilled water. Add this solution to 620 mL distilled water. Then add 130 mL 0.1N HCl.

Check the pH to be sure it is 2. Adjust with 0.1N HCl if the pH is higher than 2. Adjust with 0.1N NaOH if the pH is lower than 2. Put a caution label on the bottle.

pH 4 Buffer

Each team needs ~5 mL and an additional 80 mL per lab section.

To make 1 L:

Dissolve 10.21 g potassium hydrogen phthalate into 350 mL distilled water. Dilute up to 500 mL with distilled water and then add 499 mL distilled water. Add 1 mL 0.1N HCl.

Check the pH to be sure it is 4. Adjust with 0.1N HCl if the pH is higher than 4. Adjust with 0.1N NaOH if the pH is lower than 4.

pH 6 Buffer

Each team needs ~5 mL and an additional 80 mL per lab section.

Solutions A and B are the same for pH 6, 7, and 8 buffers.

Solution A

Dissolve 27.22 g KH_2PO_4 (potassium phosphate monobasic) in 700 mL distilled water. Bring up to 1000 mL with distilled water.

Solution B

Dissolve 34.84 g K_2HPO_4 (potassium phosphate dibasic) in 700 mL distilled water. Bring up to 1000 mL with distilled water.

To make 1 L of buffer:

Mix 877 mL Solution A with 123 mL Solution B.

Check the pH to be sure it is 6. Adjust with 0.1N HCl if the pH is higher than 6. Adjust with 0.1N NaOH if the pH is lower than 6.

pH 7 Buffer

Each team needs ~5 mL and an additional 80 mL per lab section.

Solutions A and B are the same for pH 6, 7, and 8 buffers.

Solution A

Dissolve 27.22 g KH_2PO_4 (potassium phosphate monobasic) in 700 mL distilled water. Bring up to 1000 mL with distilled water.

Solution B

Dissolve 34.84 g K_2HPO_4 (potassium phosphate dibasic) in 700 mL distilled water. Bring up to 1000 mL with distilled water.

To make 1 L of buffer:

Mix 390 mL Solution A with 610 mL Solution B.

Check the pH to be sure it is 7. Adjust with $0.1N$ HCl if the pH is higher than 7. Adjust with $0.1N$ NaOH if the pH is lower than 7.

pH 8 Buffer

Each team needs ~5 mL and an additional 80 mL per lab section.

Solutions A and B are the same for pH 6, 7, and 8 buffers.

Solution A

Dissolve 27.22 g KH_2PO_4 (potassium phosphate monobasic) in 700 mL distilled water. Bring up to 1000 mL with distilled water.

Solution B

Dissolve 34.84 g K_2HPO_4 (potassium phosphate dibasic) in 700 mL distilled water. Bring up to 1000 mL with distilled water.

To make 1 L of buffer:

Mix 53 mL Solution A with 947 mL Solution B.

Check the pH to be sure it is 8. Adjust with $0.1N$ HCl if the pH is higher than 8. Adjust with $0.1N$ NaOH if the pH is lower than 8.

pH 10 Buffer

Each team needs ~5 mL and an additional 80 mL per lab section.

To make 1 L:

Dissolve 3.96 g KCl (potassium chloride) and 3.9 g H_3BO_3 (boric acid) in 350 mL distilled water. Dilute up to 500 mL with distilled water. Add 437 mL $0.1N$ NaOH. Bring up to 1000 mL with distilled water.

Check the pH to be sure it is 10. Adjust with $0.1N$ HCl if the pH is higher than 10. Adjust with $0.1N$ NaOH if the pH is lower than 10.

pH 12 Buffer

Each team needs ~5 mL and an additional 80 mL per lab section.

To make 1 L:

Dissolve 3.55 g Na_2HPO_4 (sodium phosphate dibasic) in 350 mL distilled water. Dilute up to 700 mL with distilled water. Add 269 mL $0.1N$ NaOH. Bring up to 1000 mL with distilled water.

Check the pH to be sure it is 12. Adjust with 0.1N HCl if the pH is higher than 12. Adjust with 0.1N NaOH if the pH is lower than 12. Put a caution label on the bottle.

Phenol Red

Each section will use ~20 mL of dilute solution.

To make 1 L stock solution:

Dissolve 0.4 g phenol red in 10 mL 0.1N NaOH. Dilute up to 1000 mL with water.

Dilute stock solution 1:10 with tap water for this lab topic.

0.1M Sodium Bicarbonate, 5% Solution

Each group will use ~1 mL.

To make 1 L:

Dissolve 0.5 g $NaHCO_3$ in ~100 mL distilled water.

Summary of Materials

	Per team of students	Per room or lab section
Disposables		
cabbage (red)		¼
cheesecloth		10" × 10"
drink straw	10 mL	1
grape juice (white)	1 mL	
Maalox (diluted)	1 mL	
Milk of Magnesia (diluted)	1 mL	
Parafilm squares	20	
seltzer water	1 mL	
7UP or Sprite	1 mL	
white wine	1 mL	
Equipment		
hot plate		1
pH meter	1	
stir plate with stir bar (optional)	1	
Labware		
50-mL beakers		2
100-mL beaker	1	
250-mL beaker	1	
600-mL beaker	1	
l-L beakers		2
cutting board		1
dropping bottles	8	12
funnel (small)		1
100-mL graduated cylinder	1	
knife		1
1-mL pipets	2	
5-mL pipets	8	
1-mL pi-pump	1	
5-mL pi-pump	1	
test tubes (see discussion in text)	15	
test tube rack	1	
wash bottle	1	
wax marker	1	

Summary of Materials (Continued)

	Per team of students	Per room or lab section
Solutions/Chemicals		
pH 2 buffer	5 mL	150 mL
pH 4 buffer	5 mL	150 mL
pH 6 buffer	5 mL	150 mL
pH 7 buffer	5 mL	150 mL
pH 8 buffer	5 mL	150 mL
pH 10 buffer	5 mL	150 mL
pH 12 buffer	5 mL	150 mL
0.1N HCl*	20 mL	
phenol red (diluted solution)		40 mL
5% $NaHCO_3$	1 mL	
0.1N NaOH*	20 mL	

*Each student team that does an investigation of buffering capacity will need an additional amount of these solutions (~20 mL of each).

LAB TOPIC 4

Macromolecules

Students will be working in pairs or small groups (teams) throughout these exercises. Some of the same items are needed in more than one activity. For example, a hot plate is used in Exercise 4.1, Activity A, and in Exercise 4.4, Activities A and B. Items that can be used in different activities are marked with an asterisk (*). Solutions and disposable items used in different exercises are not marked as reusable; the amount used in each activity is given. See the Summary of Materials for the total amount.

Exercise 4.1: Carbohydrates

Activity A: Monosaccharides and Disaccharides
Materials for each student team

- Hot plate*
- Several boiling chips*
- 400-mL beaker (used as water bath)*
- 2 test tubes (regular size)
- Test tube rack*
- Test tube holder (to remove hot test tubes from water bath)*
- Potholder or insulated glove (to handle water bath)*
- Wax pencil*
- <1 mL distilled water in dropping bottle
- <1 mL 0.1M glucose in dropping bottle
- 1 mL Benedict reagent in dropping bottle (Carolina Biological Supply #84-7111 and 84-7113)

Activity B: Starch
Materials for each student team

- Test tube rack*
- Wax pencil*
- 2 test tubes (regular size)
- <1 mL 1% starch in bottle
- <1 mL distilled water in dropping bottle
- <1 mL iodine reagent (I_2KI) in brown dropping bottles (put a caution label on each bottle)

Activity C: Hydrolysis of Carbohydrates
Materials for each student team

- Hot plate*
- Several boiling chips*

- 400-mL beaker (used as water bath)*
- Test tube rack*
- Test tube holder (to remove hot test tubes from water bath)*
- Potholder or insulated glove (to handle water bath)*
- Wax pencil*
- 250-mL beaker
- Pi-pump for 5- and 10-mL pipets (green)*
- 2 10-mL pipets
- 1 5-mL pipet
- 2 extra-large test tubes
- 8 test tubes (regular size)
- 5 pasteur pipets with 1 rubber bulb
- 7-mL bottle of 2N HCl with test tube taped to side to hold pipet (put a caution label on the bottle)
- 10-mL bottle of 0.1M sucrose
- 10-mL bottle of 1% starch
- 5 mL Benedict reagent in dropping bottle (Carolina Biological Supply #84-7111 and 84-7113)
- 2 mL iodine reagent (I_2KI) in brown dropping bottles

Exercise 4.2: Lipids

Materials for each student team

- 2–3" squares of brown paper (unglazed paper such as wrapping paper or grocery bags)
- <1 mL vegetable oil in dropping bottle (any kind can be used, e.g., safflower or sunflower, also used in Lab Topic 14 but that must be light-colored)
- <1 mL distilled water in dropping bottle

Exercise 4.3: Proteins

Materials for each student team

- Wax pencil*
- 2 test tubes (regular size)
- 1 mL 1% egg albumin in dropping bottle
- Dropping bottle of distilled water
- <1 mL biuret reagent in dropping bottle (Carolina Biological Supply #84-8211 and 84-8213), diluted 1 part biuret reagent: 4 parts distilled water

Exercise 4.4: Macromolecules in Food

Activity A: Separation of Butter
Materials for each student team

- Hot plate*
- Several boiling chips*
- 400-mL beaker (used as water bath)*

- Test tube rack*
- Test tube holder (to remove hot test tubes from water bath)*
- Potholder or insulated glove (to handle water bath)*
- Wax pencil*
- 250-mL beaker*
- 10 test tubes (regular size)
- 5 pasteur pipets with 1 rubber bulb
- 50-mL beaker
- Spatula (type doesn't matter; will be used to transfer butter from the central supply into a test tube)
- Ice chest with ice: Place in a central location for all teams to share. It should be large enough to accommodate the 50-mL beaker from each group and the supply of butter.
- 2 mL biuret reagent in dropping bottle (Carolina Biological Supply #84-8211 and 84-8213), diluted 1 part biuret reagent: 4 parts distilled water
- $1\frac{1}{3}$ tablespoons butter (1 tablespoon plus one teaspoon): Get the least expensive brand, either salted or unsalted. However, stick butter must be used. Whipped butter (in tubs) will not work. And you can't substitute margarine! Place in the ice chest in a central location.
- 5 mL Benedict reagent in dropping bottle (Carolina Biological Supply #84-7111 and 84-7113)
- 1 mL iodine reagent (I_2KI) in brown dropping bottles

Activity B: Tests with Food

Materials for each student team

- Hot plate*
- Several boiling chips*
- 400-mL beaker (used as water bath)*
- Test tube rack*
- Test tube holder (to remove hot test tubes from water bath)*
- Potholder or insulated glove (to handle water bath)*
- Wax pencil*
- Pi-pump for 10-mL pipet (green)*
- 10-mL pipet
- 20 test tubes (regular size): If you need to conserve glassware, students can wash and reuse test tubes from Exercises 4.1 and 4.3 instead of getting out clean ones.
- Cutting board: We got thin acrylic panels from a hardware store and cut them into 6" squares. They are easy to clean and store, and are used for other labs in this manual.
- Single-edged razor blade
- Banana: Each team will need only ~1 cm³, but expect wastage.
- ~1 cm³ coconut: Either fresh or shredded coconut can be used.
- 5 mL milk in dropping bottle
- Raw peanut (do not use roasted or salted peanuts)
- ~1 cm³ potato
- 5 pasteur pipets with 1 rubber bulb
- 5 small squares of brown paper
- 3 mL biuret reagent in dropping bottle (Carolina Biological Supply #84-8211 and 84-8213), diluted 1 part biuret: 4 parts distilled water
- 5 mL Benedict reagent in dropping bottle (Carolina Biological Supply #84-7111 and 84-7113)
- 2 mL iodine reagent (I_2KI) (supply to lab in brown dropping bottles)
- 40-mL bottle of distilled water

Recipes

Biuret Reagent

Each team will use ~6 mL diluted reagent.

Purchase biuret reagent that has already been prepared (Carolina Biological Supply #84-8211 or 84-8213). Dilute that solution before supplying it to the labs.

To make 100 mL:

Dilute 20 mL biuret with 80 mL distilled water.

Supply to the labs in dropping bottles.

1% Egg Albumin

Each group will use ~1 mL.

To make 100 mL:

Dissolve 1 g powdered egg albumin in 100 mL distilled water.

Supply to the labs in dropping bottles.

0.1M Glucose

Each group will use <1 mL.

To make 100 mL:

Dissolve 1.8 g glucose in 60 mL distilled water. Bring the volume up to 100 mL with distilled water.

Supply to the labs in dropping bottles.

Iodine Reagent (I₂KI)

Each group will use ~6 mL.

To make 100 mL:

Dissolve 0.75 g KI (potassium iodide) in 100 mL distilled water.

Add 0.15 g I₂ (iodine) and stir to dissolve.

Store in brown bottles. Supply to the labs in *brown* dropping bottles. Put caution labels on all of the bottles.

1% Starch

Each group will use ~11 mL.

To make 100 mL:

Dissolve 1 g starch in 100 mL distilled water.

Heat to boiling to get the starch completely dissolved. The solution will initially be milky. When the starch dissolves, the solution will clear.

Supply to the labs in bottles (100–500-mL size, depending on the number of lab sections that meet in a lab room).

0.1M Sucrose

Each group will use ~10 mL.

To make 100 mL:

Dissolve 3.42 g sucrose in 60 mL distilled water.

Bring up to 100 mL with distilled water.

Supply to the labs in bottles (250–500 mL, depending on the number of lab sections that meet in a lab room).

Summary of Materials

	Per team of students	Per room or lab section
Disposables		
banana	1 cm³	
boiling chips	several	
brown paper	6 squares	
butter	1⅓ T	
coconut	1 cm³	
milk	5 mL	
peanut (raw)	1	
potato	1 cm³	
vegetable oil	1 drop	
Equipment		
hot plate	1	
Labware		
50-mL beaker	1	
250-mL beaker	1	
400-mL beaker	1	
250–500-mL bottles	4	
cutting board	1	
dropping bottle (brown)	1	
dropping bottles (clear)	4	
ice chest		1
pasteur pipets (one rubber bulb)	15	
potholder	1	
5-mL pipets	1	
10-mL pipets	3	
pi-pump for 5-mL and 10-mL pipets	1	
single-edged razor blade	1	
spatula	1	
test tubes (regular)	44	
test tubes (extra large)	2	
test tube rack	1	
test tube holder	1	
wax pencil	1	

Summary of Materials (Continued)

	Per team of students	Per room or lab section
Solutions/Chemicals		
0.1M glucose	<1 mL	
Benedict reagent	16 mL	
biuret reagent (diluted)	6 mL	
1% egg albumin	1 mL	
2N HCl	7 mL	
Iodine reagent (I$_2$KI)	6 mL	
1% starch	11 mL	
0.1M sucrose	10 mL	
water (in dropping bottle)	1 mL	
water (in larger bottle)	40 mL	

Using the Microscope

In this lab topic, students will learn how to use the microscopes they will be using for the rest of the course. Compound microscopes are used in Exercises 5.1 and 5.2. Dissecting scopes are used in Exercises 5.3 and 5.4. The maximum number of microscopes should be made available. Ideally, each student should have his or her own.

Exercise 5.1: The Compound Microscope

Activity A: The Parts of the Compound Microscope
Materials for each student or team

Only a compound microscope is needed for this part of the exercise.

Activity B: Using the Compound Microscope
Materials for each student or team

- Compound microscope
- Letter "e" slide: You can buy letter "e" slides (Carolina Biological Supply #BM 1) or make your own. To make the slides, cut out a letter "e" that has been typed on white paper using a small font size, 8 or 9 point. Place it near the center of a microscope slide and hold it down with a teasing needle while you fix it to the slide by brushing a small amount of Permount or clear fingernail polish over it. Try to get a thin, even film of mounting medium rather than a large blob.
- Transparent 6" ruler: The ruler must be transparent, since students will view it under the microscope. Plastic rulers that are 2–3 mm thick are preferable to the flat kind that are only about 1 mm thick. However, it's not worth buying a whole new set of rulers just for this exercise. If you already have the flat kind, give the students clean glass slides to put under them when they look at the rulers under the microscope.

Optional materials

The instructor may want to have students make their own letter "e" slides as wet mounts. In that case, provide each students with a letter "e," a clean slide, coverslip, and dropping bottle of water.

Activity C: Depth of Focus
Materials for each student or team

- Microscope slide and coverslip
- Cotton ball
- Forceps
- Container of *Paramecium* (Carolina Biological Supply #L 2A); can be shared among students
- Dropper for *Paramecium*
- Teasing needle

Exercise 5.2: Observing Cells

Activity A: Animal Cells

Materials for each student or team

- Toothpick: Students will use the toothpicks to scrape cells from the inside of their cheeks, so get the kind of toothpicks that have one flat, broad end.
- Container for toothpick disposal: Used toothpicks may contain biohazardous material, so an appropriate disposal container should be provided in the lab.
- Microscope slide and coverslip: Students can be expected to wash and reuse the same slides throughout the week if you have multiple lab sections.
- Staining stations: Students can share the staining stations; set up one station for every 4 or 5 students.
- Small Petri dish half and large Petri dish half: Place the small Petri dish half (60-mm size) inside the large Petri dish half (150-mm size). The small dish simply serves as the platform for resting the slide, while the large dish is a catch basin for excess stain and water. Any other pieces of labware that serve this purpose are fine.
- Bottle of methylene blue, 2 mL per student (see Recipes section)
- Container for disposal of waste stain
- Pasteur pipet with rubber bulb (for dispensing methylene blue): Tape a test tube to the bottle and keep the pipet in that.
- Wash bottle of tap water
- Dropping bottle of tap water
- Box of cleaning tissues
- Prepared slide of skin: This should be a cross section (e.g., Carolina Biological Supply #H 7475), but the exact type of skin is not important. Slides can also be shared among students.

Activity B: Plant Cells

Materials for each student or team

- Microscope slide and coverslip: The slides can be washed and reused by students throughout the week.
- *Elodea* leaf: Each student will use only one *Elodea* leaf, so you don't need large amounts. The *Elodea* should be supplied to the labs in water that is room temperature or a little warmer for the best results. If you plan to take the *Elodea* directly from a refrigerator or coldroom to the labs, rinse it in warm running water first.

Optional Materials

The instructor may want you to supply extra specimens for microscope practice. Pond or aquarium water is a good source of microorganisms. You can also maintain a culture of microorganisms by making a hay infusion (see Recipes section for one possible recipe).

Exercise 5.3: The Stereoscopic Dissecting Microscope

Activity A: The Parts of the Dissecting Microscope

Materials for each student team

Students will only need the dissecting scope (stereomicroscope) for this activity.

Activity B: Using the Dissecting Microscope

Materials for each student team

- "e" slide: Use the same "e" slides used in Exercise 5.1, Activity B.
- Photo from newspaper: Should be small enough to set on the stage of the dissecting scope, but it makes no difference what the pictures are of, nor do they all need to be similar.

Exercise 5.4: Observing an Animal Specimen

Materials for each student team

- Prepared slide of hydra: This should be a whole mount. Carolina Biological Supply #Z 605 is good, but if you already have other whole mounts of hydra they will do just as well.
- Living hydra in small dish: For the living specimens of hydra, just get the cheapest kind available (e.g., Carolina Biological Supply L55). It is most convenient to supply them in small Petri dishes or watch glasses so that students don't have to transfer them from one container to another in order to view them on the dissecting scope.
- Container of beef liver juice and piece of thread: Each team will dip a short piece of thread in liver juice to stimulate the hydra's feeding response. The threads should be disposed of after use. Make the liver juice by mashing a little bit of beef liver in some of the juice from the container. Each class will require a very small amount of the juice and it will go bad quickly, so supply it in very small quantities.

Optional Materials

The instructor may ask you to provide additional specimens for students to view with the dissecting scope.

Recipes

Methylene Blue

Each student or group will use ~2 mL.

To make 1 L:

Dissolve 1 g methylene blue powder in 1000 mL of 100% ethanol.

Provide waste containers in the labs for excess stain.

Hay Infusion

Fill a large jar or beaker (1 or 2 L) one-quarter full of pond water. Add a tablespoon of mud from the bottom of a pond, some dried leaves, and a handful of hay (dried timothy grass or other dried weeds can be substituted). Put the jar, uncovered, in a warm place for a few days, and then add 6 grains of uncooked rice. The plant material will decay and drop to the bottom, the water will become cloudy, and there may be a scum on the surface. It takes at least 2 weeks to get a good culture started.

Summary of Materials

	Per team of students	Per room or lab section
Disposables		
cotton ball		1
beef liver juice		~10 mL
thread	~2"	
toothpick	1	
Equipment		
compound microscope	1	
dissecting microscope	1	
Labware		
dropper	1	
forceps	1	
pasteur pipet	1	
Petri dish (small)	1	
Petri dish (large)	1	
photo (newspaper)	1	
ruler (transparent)	1	
slides and coverslips	3	
teasing needle	1	
wash bottle	1	
waste stain container		1
Fresh Materials		
Elodea	1 leaf	
hydra	1	
Paramecium	1 drop	
Prepared Slides		
letter "e"	1	
hydra	1	
skin section	1	
Solutions/Chemicals		
methylene blue	~2 mL	

Enzymes

In this lab topic, students learn a method for investigating an enzyme reaction (Exercise 6.2). In Exercise 6.3, students design an experiment using this method. In Exercise 6.4 they perform their own experiment. Some instructors may do Exercises 6.1, 6.2, and 6.3 one week and Exercise 6.4 the following week.

Students will work in teams throughout the exercises. The instructor will specify the number of students per team (usually 2–4) and tell you how many teams to prepare for per lab section. The quantities of materials needed by each team are the same regardless of team size.

Exercise 6.1: Factors Affecting Reaction Rate

No prep is required for this exercise.

Exercise 6.2: The Catecholase-Catalyzed Reaction

Materials for the instructor

- White potato: ~80 g (about half of a small potato). Any variety can be used. Put out whole potatoes and let the lab instructors cut them as needed.
- Blender (to grind potatoes)
- 600-mL beaker
- Cheesecloth square, 2 layers: Approximately 8" × 8", used to strain potato extract.
- Balance with large weigh boat
- Knife
- Cutting board
- 500-mL graduated cylinder

Materials for each student team

- Spectronic 20: Color chart method can be substituted if Spec 20s or other spectrophotometers are not available. Colorimeters can also be used. For this experiment, use a green filter.
- 4 small test tubes or cuvettes for spectrophotometer
- Test tube rack
- 1 1-mL pipet: The pipets can be labeled so that they can be reused by successive lab sections. The 1-mL pipet is for potato extract.
- Pi-pump for 1-mL pipet (blue)
- 2 5-mL pipets (for water and catechol)
- Pi-pump for 5-mL pipet (green)
- 5 mL 0.05% catechol in brown bottle (keep in dark bottle)
- 10 mL distilled water in bottle

- 4 small Parafilm squares: Used to cover the test tubes. Cut standard (2") Parafilm squares into quarters.
- Vial with cap that fits tightly, to hold up to 50 mL (e.g., snap-cap bottles, Carolina Biological Supply #71-5012)
- Wax pencil

Exercise 6.3: Designing an Experiment

No materials are needed for this exercise.

Exercise 6.4: Performing the Experiment and Interpreting the Results

In this lab topic, students are given a method for assaying catecholase (Exercise 6.2) and asked to design an experiment using this method. In Exercise 6.4 they perform their own experiment. Each student team should receive a basic kit for performing the assay, which includes many of the materials supplied for Exercise 6.2. Prepare as many basic kits as will be needed in one lab section and put them in the room in boxes, plastic dishpans, or some other container. Students can be expected to clean their glassware and reassemble the kits when they finish lab, so the kits can be reused through multiple lab sections.

Basic Kit for Catecholase

Items from Exercise 6.2 that students can reuse are marked with an asterisk (*).

- Vial with cap that fits tightly, to hold up to 50 mL*
- 1-mL pipets labeled "potato extract"*
- 2 5-mL pipets labeled "catechol" and "water"*
- Pi-pump for 1-mL pipet*
- Pi-pump for 5-mL pipet*
- Wax pencil*
- Test tube rack*
- Bottle of distilled water*
- 20 small test tubes or Spec 20 tubes
- 20 small Parafilm squares
- 25 mL catechol (0.05%) in dark bottle

Additional Materials

Each student team will also need additional materials. Consult with the instructor regarding materials and quantities that you should have available. Additional materials for possible student investigations are given below, along with approximate quantities per team. These items can be placed in a central location.

Concentration of enzyme or substrate

No additional materials are required.

pH

- pH buffers, ~10–15 mL of each pH: Recipes for pH 2, 4, 6, 7, 8, 10, and 12 buffers are in the Recipes section of Lab Topic 3. Put caution labels on bottles of pH 2 and pH 12 buffers.

Temperature

- Water baths, if available
- Hot plate
- Ice
- Thermometer

Salt concentration

- Salt (probably NaCl) solutions in a range of concentrations, for example, 0.5, 0.9, 1.0, 1.5, 2.0, and 5.0%: Students will use 10–15 mL of each concentration. If the instructor doesn't specify concentrations to have available, provide a jar of dry NaCl or other salt, balance, weigh boat, spatula, 100-mL graduated cylinder, stir rod, and beakers or flasks so students can mix their own solutions.

Source of enzyme

Consult with the instructor about what to purchase. Students might be interested in using any fruit or vegetable that shows a browning reaction, including apples, pears, bananas, or different varieties of white potatoes. Students will make an extract using the same materials used by the instructor for Exercise 6.1. Approximately 20 g of each source should be sufficient for each team that chooses this investigation.

Pretreatment

If the schedule permits, students may treat the potatoes in some fashion before extracting the enzyme. For example, they may store potatoes under different conditions of temperature or humidity for a week. When they perform their experiment, students will use the materials supplied for the instructor to make a potato extract in Exercise 6.1.

Inhibition by ascorbic acid

To test inhibition of browning by citrus juice, ascorbic acid, or a product such as Fruit-Fresh, students will need approximately 10 mL of the inhibitor. Concentrations of ascorbic acid might range from 0.1% to 5% if students want to investigate the effect of concentration.

Recipe

Catechol, 0.05%

To make 1 L:
Dissolve 0.5 g catechol in 1000 mL distilled water.

Keep solution in a brown bottle.

Summary of Materials

	Per team of students	Per room or lab section
Disposables		
cheesecloth, 2 layers		1 square
Parafilm (small squares)	4	
potato (white)		~80 g
Equipment		
balance with weigh boat		1
blender		1
Spec 20 or other spectrophotometer	1	
Labware		
600-mL beaker		1
250-mL bottle	1	
cutting board		1
500-mL graduated cylinder		1
1-mL pipet	1	
5-mL pipet	2	
1-mL pi-pump	1	
5-mL pi-pump	1	
cleaning tissues (if Spec 20 used)	1 box	
knife		1
small test tubes (cuvets if using Spec 20)	4	
test tube rack	1	
vial with cap	1	
wax pencil	1	
Solutions/Chemicals		
catechol, 0.05%	5 mL	
distilled water	10 mL	

Cellular Respiration

✳ NOTE: Lima beans for Exercise 7.2 must be soaked overnight.

In this lab topic, students learn a method for investigating alcoholic fermentation (Exercise 7.1) and a method for investigating aerobic respiration (Exercise 7.2). In Exercise 7.3, students design an experiment using one of these methods. In Exercise 7.4 they perform their own experiment.

Students will work in teams throughout the exercises. The instructor will specify the number of students per team (usually 2–4) and tell you how many teams to prepare for per lab section. The quantities of materials needed by each team are the same regardless of team size.

Exercise 7.1: Alcoholic Fermentation

Materials for each student team

- 3 fermentation set-ups, consisting of the following items: 2 large test tubes (25 × 150 mm, e.g., Carolina #73-1024) or 8 dram shell vials (Carolina #71-5126), a #4 one-hole rubber stopper (fits large test tube; size may be different for shell vial), a 400- or 600-mL beaker and a 15" piece of Tygon tubing (must fit hole in rubber stopper, e.g., Carolina #71-1602; the tubing sold for aquarium pumps also works well). The general configuration of the fermentation kit is shown in Figure 7.1. Other types of vials may be used instead of the test tubes, but their diameter should be approximately 2.5 cm to give good results with the procedure as written. Graduated test tubes can be used for one of the test tubes in each kit, but this is not recommended because of the cost (approximately $20 apiece).
- 250-mL beaker or a test tube rack large enough to accommodate extra-large test tubes while students are mixing solutions for the experiment
- Wax marker
- Bottle of water: Each group will use ~10 mL (tap water is okay). It can be put out in a large bottle to be used by successive lab periods.
- Bottle of corn syrup, diluted 1:1 (each group will use ~10 mL): To ensure good results, use Karo light corn syrup. Other brands may contain a lower concentration of sugar (especially anything called "lite") or include preservatives or other chemicals that affect results. Dilute the syrup 1:1 with tap water before putting it out in the labs.
- Test tube for yeast suspension
- 15 cm (6") ruler (must have metric markings)

Figure 7.1
Fermentation apparatus.

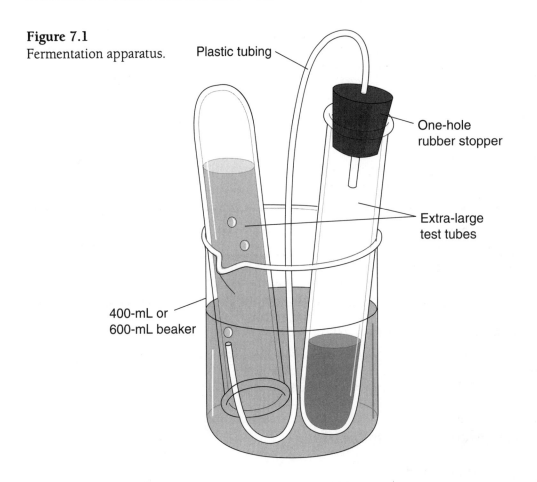

- 3 5-mL pipets (can be reused by different classes if you label them for use with water, syrup, and yeast)
- Pi-pump for 5-mL pipet

Materials for the instructor

- 1 package of Rapid Rise yeast per 6 student teams: Use Fleischmann's Rapid Rise yeast. If you must use regular yeast instead, you will need twice as much. The lab instructors will make up a fresh yeast suspension (1 package Rapid Rise in 50 mL water) before each lab. Leftover yeast suspension can be saved for the next lab period but should not be kept overnight.
- 250-mL beaker
- Stir rod

Materials to have available by the sink

- 1 or 2 dishtubs: Place them at sinks to be shared by student groups. If dishtubs are not available, have a sink stopper so that the sink can be filled with water.

Exercise 7.2: Aerobic Respiration

✳ NOTE: Lima beans must be soaked the night before they are needed.

Consult with the course instructor about who will make up the suspension and when (we have our lab instructors do it). The procedure takes about 20 minutes. Instructions are also printed in the Annotated Instructor's Edition of the lab manual. If the suspension is kept on ice, it will be usable for about 5–6 hours, but it shouldn't be kept longer than that. One batch yields ~50 mL.

Materials needed to prepare the mitochondrial suspension

- 50 g baby white lima beans, *soaked overnight:* Use dried baby white lima beans for the mitochondrial suspension. Buy beans at the grocery store. Don't get seed beans, which have often been treated with fungicides. Don't use green-colored beans of any type, since the chloroplasts they contain will interfere with the results of the experiment.
- Tabletop centrifuge
- Ice chest with ice

✳ NOTE: Place all glassware in the freezer several hours before making the suspension. The solutions and cheesecloth should be refrigerated.

- 600-mL beaker (place in freezer prior to use)
- 150-mL beaker (place in freezer prior to use)
- 100-mL grad cylinder (place in freezer prior to use)
- Blender (keep cup in freezer)
- 6 centrifuge tubes in test tube rack (place in freezer prior to use)
- Several squares of 8" × 8" cheesecloth (place in refrigerator prior to use)
- Sucrose phosphate buffer (~100 mL per suspension; place in refrigerator prior to use); see Recipes section

Procedure for preparing the mitochondrial suspension

Weigh out 50 g of dried lima beans and soak them in water overnight before the suspension is made. Put the beans in plenty of water (a 250-mL beaker works well for 50 g) since they will take up water, and keep them refrigerated.

All solutions and containers should be kept ice cold! Place the blender cup and glassware that you will be using in the freezer several hours before you begin the procedure, and keep all of the solutions refrigerated.

1. Place 50 g dried lima beans (soaked overnight) in blender with 100 mL sucrose-phosphate buffer. Blend at high speed for 2 minutes.
2. Set a 600-mL beaker in an ice bucket. Strain the homogenate through cheesecloth into this beaker. You may have to change the cheesecloth once or twice as it becomes clogged with particles. Squeeze the cheesecloth gently to get the filtrate through.
3. Divide the filtrate equally among the centrifuge tubes (there will be ~10 mL per tube).
4. Centrifuge at 200 × g (or at a low setting) for 1 minute. Do not decant.
5. Increase the speed and centrifuge at 1300 × g (or at a medium-high setting) for 10 minutes.
6. Decant the supernatant from each tube into a beaker and put the beaker in an ice bucket. This suspension contains mitochondria. Supply it to the labs in small test tubes. Each group will use only about 1 mL, so put out two or three test tubes to be shared by groups. The mitochondrial suspension should be kept on ice in the lab.

✳ NOTE: Many tabletop centrifuge models do not indicate × g settings. You can obtain this information from the manufacturer. If you can't determine what the settings mean, use a low speed for the first run, and then increase to a medium to high speed. Keep the pellets. If you centrifuged at too high a speed, the mitochondria will be in the pellet, which can then be resuspended in sucrose-phosphate buffer.

Materials for each student team

- Spec 20: If Spec 20s are not available, the experiment can be done using the color chart in the Lab Investigations manual (see Color Plate 1).
- 3 small test tubes (or cuvets): We use ordinary small test tubes (13 × 100 mm) in the Spec 20. None of the experiments in these lab topics are sensitive enough to require the more expensive cuvets. However, do not use test tubes that are scratched or have been etched by dish soap.
- Test tube rack
- Wax marker
- Bottle of phosphate buffer (~15 mL/team; keep on ice; see Recipes section)
- Bottle of 0.025M succinate (< 1 mL/team; see Recipes section)
- Bottle of 1 mM DCPIP (~1 mL/team; see Recipes section)
- 5-mL pipet
- 3 1-mL pipets: The pipets can be reused by successive lab sections if you label them for DCPIP (1 mL), succinate (1 mL), mitochondrial suspension (1 mL), and buffer (5 mL). Tape a test tube to the side of each stock bottle and keep the pipet for each bottle in its test tube.
- Pi-pump for 5-mL pipet
- Pi-pump for 1-mL pipet
- 3 small squares of Parafilm: Used to cover the small test tubes. Cut the regular 2" squares into quarters.
- Box of cleaning tissues

Exercise 7.3: Designing an Experiment

No materials are needed for this exercise.

Exercise 7.4: Performing the Experiment and Interpreting the Results

Each student team should receive a basic kit for performing *either* the method in Exercise 7.1 *or* the method in Exercise 7.2. The basic kit includes some materials students have already used for Exercises 7.1 and 7.2. Consult with the instructor about the number of teams that will need each kit. Prepare as many basic kits as will be needed for one lab section and put them in the lab room in boxes (or plastic dishpans or some other container). Students can be expected to clean their glassware and reassemble the kits when they are finished, so kits can be reused through multiple lab sections.

Basic Kit for Alcoholic Fermentation (Exercise 7.1)

Items from Exercise 7.1 that students can reuse are marked with an asterisk (*). Quantities of solutions are in addition to the quantities needed for Exercise 7.1.

- 10 fermentation setups, each consisting of the following (see description of materials in Exercise 7.1). Each team already has 3 setups from Exercise 7.1. Students can also obtain setups from teams who are designing investigations based on Exercise 7.2 rather than Exercise 7.1:

     2 large test tubes

     1 #4 one-hole rubber stopper

     1 400- or 600-mL beaker

     1 15" length of Tygon tubing
- 250-mL beaker or test tube rack to hold large test tubes*
- Wax marker*
- Bottle of water*
- Bottle of corn syrup, diluted 1:1 (~50 mL)*
- 6" ruler*
- 3 5-mL pipets—labeled "water," "syrup," and "yeast"*
- Pi-pump for 5 mL pipet*
- Package of Rapid Rise yeast (or 2 packages if regular yeast is used)
- 150-mL beaker
- 1 stir rod

Materials to have available by the sink

- 1 or 2 dishtubs

Basic Kit for Aerobic Respiration (Exercise 7.2)

Items from Exercise 7.2 that students can reuse are marked with an asterisk (*). Quantities of solutions are in addition to the quantities needed for Exercise 7.2.

- Spec 20 (color chart may be used instead)*
- Box of cleaning tissues*
- 20 small test tubes (students can wash and reuse the test tubes used in Exercise 7.2)*
- Test tube rack*
- Wax marker*
- Bottle of phosphate buffer (~100 mL/group; keep on ice)*
- Bottle of 0.025M succinate (~25 mL/group)*
- Bottle of 1 mM DCPIP (~25 mL/group)*
- 5-mL pipet labeled "buffer"*
- 3 1-mL pipets labeled "DCPIP," "succinate," and "mitochondrial suspension"*
- Pi-pump for 5-mL pipet*
- Pi-pump for 1-mL pipet*
- Mitochondrial suspension (~25 mL/group)*
- 20 small squares of Parafilm

Additional Materials

Each student team will also need additional materials. Consult with the instructor regarding materials and quantities that should be available. Students may bring in some materials themselves. Additional materials for possible student investigations are given below, along with approximate quantities needed per team (each team will do only one investigation). These items can be placed in a central location in the lab room.

Suggested Materials for Student Investigations of Alcoholic Fermentation

Concentration of yeast (enzymes) or corn syrup (substrate)

No additional materials are required.

Different substrates

Investigations might use different brands of corn syrup, different types of sweeteners (e.g., molasses, honey, sorghum, artificial sweeteners, maple syrup), beverages that contain sugar (e.g., fruit juices and sports drinks such as 10-K), or different types of sugars (e.g., glucose, sucrose, fructose, maltose, lactose). If the latter are used, either make up 20% solutions or give students the dry chemicals and the materials needed to make up solutions themselves (balance, weigh boat, spatula, stir plate, beakers). Students could also puree whole fruits in a blender.

Each team will need approximately 15–20 mL of each substrate used.

Students will also need measuring devices such as pipets or 10-mL graduated cylinders.

Different types of yeast

Brewer's yeasts are different strains from the baker's yeast used for Exercise 7.1. Supply stores for home brewers can provide various types of yeast to compare. Several sources are listed in the Vendor Addresses section. The brewer's yeast that is sold in nutrition stores, however, cannot be used since it is not living. (It is the residue of dead yeast cells that remain after fermentation has taken place.)

Inhibition of fermentation by alcohol

Provide approximately 15 mL of 100% or 95% ethanol, and an extra 5-mL pipet for measuring.

pH

- pH buffers, approximately 20 mL of each pH: Recipes for pH 2, 4, 6, 7, 8, 10, and 12 buffers are in the Recipes section of Lab Topic 3. Provide a 5-mL pipet for each buffer. Put caution labels on the bottles of pH 2 and pH 12 buffers.

Temperature

Water baths are ideal, but students can mix the temperatures they want in dish tubs instead. Also provide ice, a hot plate, a 2-L beaker for heating water, and rubber gloves for handling fermentation apparatus in hot water. Students also need a thermometer.

Suggested Materials for Student Investigations of Aerobic Respiration

Concentration of mitochondrial suspension (enzyme) or succinate (substrate)

Students may need glassware for making dilutions, for example, extra 1-mL and 5-mL pipets, a 10-mL graduated cylinder, and test tubes.

Mitochondria from different sources

Students may extract mitochondria from other dried beans. Beans will have to be soaked overnight. All of the materials listed in Exercise 7.2 for making the mitochondrial suspension will be needed.

Inhibition

This will not be a common type of investigation, so be sure to ask the instructor if it is likely to be used. Provide 0.1M solutions of malonate and oxaloacetate. Provide a 100μg/L solution of mercuric chloride.

> ⚠ **CAUTION:** Mercuric chloride is toxic. Use caution when handling it, and be sure to label the bottle.

pH

Supply pH buffers, ~10 mL of each pH. Recipes for pH 2, 4, 6, 7, 8, 10, and 12 buffers are in the Recipes section of Lab Topic 3. Provide a 1-mL pipet for each buffer. Put caution labels on the bottles of pH 2 and pH 12 buffers.

Temperature

Water baths are ideal, but students can mix the temperatures they want in dish tubs instead. Also provide ice, a hot plate, a 2-L beaker for heating water, and rubber gloves for handling fermentation apparatus in hot water. Students also need a thermometer.

Recipes

DCPIP (dichlorophenol-indophenol), 1mM

Each team will need ~1 mL for Exercise 7.2. An additional 25 mL is needed for each basic kit.

To make 1 L:

Dissolve 0.29 g DCPIP *completely* in 100 mL 95% ethanol. You must stir the solution for at least 15–30 minutes to dissolve the DCPIP.

Bring volume up to 1000 mL with distilled water.

Solution should be kept in a dark bottle.

Make fresh solution shortly before this lab topic will be performed each quarter or semester. It shouldn't be necessary to make a fresh solution before each lab period, but make it up during the week before Lab Topic 7 will be performed. It may be necessary to stir the solution again thoroughly before placing it in the labs. Discard any leftover solution; don't save for following term.

DCPIP (also called DPIP) is 2,6-dichlorophenol-indophenol. It is also known as 2,6-dichloro-indophenol. The powdered form can be ordered from Sigma (#D1878), Fisher Scientific (#S286-5), or Carolina Biological Supply (#86-8600), as well as other chemical supply houses.

Phosphate Buffer

Each team will need ~15 mL for Exercise 7.2. An additional 100 mL is needed for each basic kit.

To make 1 L:

Solution A (0.2M KH₂PO₄):

Dissolve 27.22 g KH_2PO_4 (monobasic potassium phosphate anhydrous) in 600 mL distilled water.

Bring volume up to 1000 mL with distilled water.

Solution B (0.2M K₂HPO₄):

Dissolve 34.84 g K_2HPO_4 (dibasic potassium phosphate anhydrous) in 600 mL distilled water.

Bring volume up to 1000 mL with distilled water.

For 1 L of buffer, mix 280 mL Solution A with 720 mL Solution B. The pH should be 7.2. Adjust by adding Solution A if pH is too high or Solution B if pH is too low.

Succinate (0.025M)

Each team will need 1 mL for Exercise 7.2. An additional 25 mL is needed for each basic kit.

To make 1 L:

Dissolve 2.59 g succinic acid in 600 mL distilled water. Bring volume up to 1000 mL with distilled water.

Sucrose-Phosphate Buffer

100 mL needed for each mitochondrial extraction

To make 1 L:

Dissolve 136.9 g sucrose in phosphate buffer, pH 7.2 (recipe above).

Check pH; adjust as described above if necessary (pH should be 7.2).

Summary of Materials

	Per team of students	Per room or lab section
Disposables		
cheesecloth		4–5 squares
cleaning tissues	1 box	
Karo light corn syrup (diluted 1:1)	10–60 mL*	
dried lima beans		50 g
Parafilm (small squares)	4–20*	
dried yeast (Fleischmann's Rapid Rise)	1 pkg*	1 pkg
Equipment		
blender		1
tabletop centrifuge		1
ice chest		1
Spec 20 or other spectrophotometer	1	
Labware		
150-mL beaker	1*	
250-mL beaker	1	1
400-mL or 600-mL beakers	3–10*	
600-mL beaker		1
100–250-mL (clear) bottles	4	
100–250-mL (brown) bottle	1	
centrifuge tubes		6
dish tubs		1–3
100-mL graduated cylinder		1
1-mL pipets	3	
5-mL pipets	4	
pi-pump (for 1-mL pipet)	1	
pi-pump (for 5-mL pipet)	1	
rubber stoppers, 1 hole (#4)	3–10	
15-cm (6") ruler	1	
stir rod	1*	
test tubes, small (or cuvets)	4–20*	
test tubes, large	6–20*	
test tube rack	1	
15" piece of Tygon tubing	3–10*	
wax marker	1	
Solutions/Chemicals		
DCPIP, 1 mM	1–26 mL*	
Phosphate buffer (7.2)	15–115 mL*	
Succinate, 0.025M	1–26 mL*	
Sucrose-phosphate buffer		100 mL

*Quantity depends on whether team will use basic kit for this method as well as perfoming Exercise 7.1 or 7.2.

LAB TOPIC 8

Photosynthesis

In this lab topic students learn a method for investigating photosynthesis (Exercise 8.1). In Exercise 8.2, students design their own experiment using this method. In Exercise 8.3 they perform their own experiment.

Students will work in teams throughout the exercises. The instructor will specify the number of students per team (usually 2–4) and will tell you how many teams to prepare for per lab section. The quantities of materials needed by each team are the same regardless of team size.

Exercise 8.1: Measuring Photosynthetic Rate in Spinach Leaf Disks

Materials for each student team

- Reflector lamp with 150W bulb: Can be purchased at a hardware store. The lamp consists of a simple socket with a metal shade and often comes with a clamp attached. Be sure to buy lamps that are rated for 150W or more. The bulbs, which are called indoor floodlights, are wider and flatter than the standard light bulb and have a reflective coating.
- Support stand with 3-prong clamp (e.g., Fisher #05-769 utility clamp)
- 1 1-L or 2-L beaker
- 0.2% $NaHCO_3$ (~300 mL/team; see Recipes section)
- 30-cm (12") ruler (must have metric markings)
- 250-mL filtering flask with side hose connection and 1-hole rubber stopper (#6): Cover hole with short piece of label tape that is folded under on one end so it can be easily peeled back; *or* you can substitute a 250-mL filtering flask (heavy wall) with 2-hole rubber stopper. Cover one hole with label tape as described above. Put a short (5–7 cm) piece of glass tubing through the other hole.
- Dish that holds at least 100 mL of fluid (e.g., culture dish #74-1004 from Carolina Biological Supply)
- #3 cork borer and glass or wooden rod to fit through it: Any metal cylinder approximately 6–8 mm in diameter, reasonably sharp on one end, may be used in place of the cork borer. A metal shop may be able to make you a set of cork borer substitutes. The rod is used to push out leaf disks after they have been cut. Pieces of dowel rod may be used for this purpose.
- Cutting board or smooth piece of Styrofoam for students to cut disks on (paper towels can be used)
- Forceps
- 3 Petri dishes
- Several *fresh* spinach leaves: Most fresh spinach will be OK, whether packaged or in loose bunches, but don't use yellow or brown leaves. And, of course, you cannot substitute canned or frozen spinach.
- Water aspirator or another vacuum source: One vacuum source can shared by an entire lab section, but it helps to have two since most students will need to use it at the same time. The

Fisher Scientific airejector aspirator (#09-956) is expensive ($35) but very effective. Use plumber's tape when you attach it to the faucet. Get a splashguard also (Splashgon cylinder, #09-959; $8.50). We have used the Nalgene vacuum pump from Fisher (#09-960-2; $5.85), but the water pressure in our sinks is sometimes very low and we had trouble getting a consistently strong enough vacuum to infiltrate the disks. Other vacuum sources available to you (e.g., a vacuum line or vacuum pump) will also serve the purpose. Carolina Biological Supply sells a hand-held vacuum pump (#75-2810); I am not familiar with how well these will work for this experiment.

- Approximately 1.5–2 feet of vacuum tubing for each aspirator or other vacuum source

Exercise 8.2: Designing an Experiment

No materials are needed for this exercise.

Exercise 8.3: Performing the Experiment and Interpreting the Results

Each student team should receive a basic kit for performing the method introduced in Exercise 8.1. The basic kit contains some materials students have already used for Exercise 8.1. Prepare as many basic kits as will be needed for one lab section and put them in the lab room in boxes (or plastic dishpans or some other container). Students can be expected to clean their glassware and reassemble the kits when they are finished, so kits can be reused by students through multiple lab sections.

Basic Kit for Photosynthesis

Items from Exercise 8.1 that students can reuse are marked with an asterisk (*). Quantities of disposables and solutions are in addition to the quantities needed for Exercise 8.1.

- Fresh spinach (several leaves)
- 1 L 0.2% $NaHCO_3$ (sodium bicarbonate): If you have numerous lab sections, it is most efficient to put a carboy of $NaHCO_3$ in the lab and let students use that instead of filling individual bottles.
- 250-mL flask with 2-hole rubber stopper*
- Water aspirator or other vacuum source with vacuum hose attached to glass tube in rubber stopper (may be shared by several teams)*
- #3 cork borer or substitute and glass or wooden rod to fit through it*
- Cutting board or smooth piece of Styrofoam for students to cut disks on (paper towels may be substituted)*
- Forceps*
- 3 Petri dishes (in addition to the 3 used in Exercise 8.1)
- Reflector lamps (at least 2, or as many as are available, in addition to the one used in Exercise 8.1)
- Support stands (one for each lamp, each with a 3-prong clamp)
- 1-L or 2-L beakers (one for each lamp)
- Culture dish that holds at least 100 mL*

Additional Materials

Each student team will also need additional materials for performing their experiments. Consult with the instructor regarding materials and quantities to have available. Students may bring in some materials themselves. Additional materials for possible student investigations are listed below, along with approximate quantities needed per team. These items can be placed in a central location in the lab room.

Concentrations of sodium bicarbonate (NaHCO₃)

Supply a range of concentrations of $NaHCO_3$ (e.g., from 0.1% to 5.0%). Since students use 0.2% $NaHCO_3$ in Exercise 8.1, they should vary the concentration in that neighborhood. Each team needs approximately 350 mL of each concentration. Alternatively, supply students with a 10% solution and have them make their own dilutions. In that case, they will need glassware for measuring and mixing, too. In a large course, it is most efficient to put carboys of 10% solution in the labs and let students mix their own. Or students can be given dry $NaHCO_3$ and equipment to weigh and mix their own solutions.

Different types of leaves

Students may use the method introduced in Exercise 8.1 to investigate photosynthesis in leaves other than spinach. For example, different types of greens (turnip, collard, cabbage, mustard) can be used. Houseplants or leaves from local trees, shrubs, or herbaceous plants can be used. We usually ask students to supply field-collected leaves, but will get them whatever is available as produce in the grocery store.

Light intensity

Light intensity can be varied by using gauze or cheesecloth, or greenhouse shadecloth if that is available. Or provide light bulbs of different wattages for the lamps. A light meter is useful but not required.

Colored filters

To vary light quality (color), students will place colored filters between the lamp and the spinach. The idea is to allow only one particular color of light to reach the plant. For example, a green filter only lets green light through. Unfortunately, not all colored acetates or cellophanes can be used for this type of experiment. Many colored materials allow several colors to pass. Almost any yellow cellophane you see, for example, transmits other colors besides yellow. You can determine which materials are good to use by testing them in a Spec 20. The following usable acrylic filters are available in 20" × 24" sheets from Edmund Scientific Company: medium green (G82,041), medium red (G82,015), and daylight blue (G82,031).

pH

Supply pH buffers, approximately 350 mL of each for each team that performs this type of investigation. Recipes for pH 2, 4, 6, 7, 8, 10, and 12 buffers are in Lab Topic 3 in this prep guide. Add 2g $NaHCO_3$ to each liter of buffer, and then recheck the pH and adjust as needed. Put caution labels on bottles of pH 2 and pH 12 buffer.

Recipe

0.2% NaHCO₃ (sodium bicarbonate)

Each student team needs approximately 1.3 L.

To make 1 L:
Dissolve 2 g NaHCO₃ (sodium bicarbonate) in 1000 mL distilled water.

Summary of Materials

	Per team of students	Per room or lab section
Disposables		
fresh spinach	~10 leaves	
Equipment		
vacuum source with vacuum tubing		1 or more
Labware		
1-L or 2-L beakers	3	
3-prong clamps	3	
cork borer or substitute	1	
cutting board	1	
dish	1	
filter flask with 1-hole stopper *or* 250-mL flask with 2-hole stopper	1	
forceps	1	
Petri dishes	6	
reflector lamps with bulb	3	
30-cm (12") ruler	1	
support stands	3	
Solutions/Chemicals		
0.2% NaHCO₃	1.3 L	

LAB TOPIC 9

Mendelian Genetics

Exercise 9.1: Mendel's Principles

No materials are needed for this exercise.

Exercise 9.2: A Dihybrid Cross Using Wisconsin Fast Plants™

This experiment takes approximately 7 weeks to complete. The schedule given here and in the Lab Investigations manual assumes that students will begin the experiment during their regular lab time 7 weeks before Lab Topic 9 is scheduled. Subsequent events are also scheduled during lab time when possible.

Wisconsin Fast Plants are available in kits from Carolina Biological Supply. It is strongly recommended that you purchase the kits the first time you do this experiment. A kit contains enough materials for 16 lab teams. Each team plants one quad. Check with the instructor about the number of lab teams per section. Refill kits and seeds are available from Carolina Biological Supply for subsequent terms. Fast Plants are indexed under W for Wisconsin Fast Plants in the Carolina catalog. The experiment described in this lab topic is the dihybrid short form genetics kit, #15-8774. Alternative materials are discussed at the end of this chapter.

A light bank is essential for growing Fast Plants, which require *constant* high light to complete their life cycle in the time allotted. Light available from a windowsill, greenhouse, or many types of growth shelves is not sufficient. In addition, students need to have access to their plants for Activity C. See details below.

Activity A: Planting F_1 Seeds
7 Weeks Before Lab Topic 9 Is Scheduled
Materials for each student team (provided in Fast Plants kit):
- 1 quad
- 4 wicks
- Potting mix
- 12 fertilizer pellets
- 1 pot label
- 1 plastic pipet for watering
- 12 F_1 seeds (ANL/anl, YGR/ygr)

Materials for each student team (not provided in kit)
- Waterproof marker

Materials for entire class (provided in Fast Plants kit)

- Watering trays with mats (each tray holds 8 quads; kit contains 2 trays)
- Copper sulfate square

Materials for entire class (not provided in Fast Plants kit)

- Light bank: Fast Plants require constant high light in order to develop, flower, and set seed according to the recommended schedule. A light bank of six 4-foot (40-watt) fluorescent (cool white, not grow lamp) bulbs should be used. The Growing Instructions booklet provided with each Fast Plants kit contains plans for building a light bank and a rack to hold it. A light system that includes everything but the light bulbs is available from Carolina Biological Supply (#15-8998). Six watering trays, each holding 8 quads, will fit under one light bank. The lights will have to be raised as the plants grow taller, and they should always be 5–8 cm from the tops of the plants.

Activity B: First Observations

6 Weeks Before Lab Topic 9 Is Scheduled

Materials for each student team (provided in Fast Plants kit)

- 1 dead bee
- Small wooden stakes and plastic rings (will be used to stake up plants as they grow taller)

Materials for each student team (not provided in kit)

- Forceps
- Toothpick
- Fast-drying glue (for example, Krazy Glue)
- Pieces of Styrofoam for students to stick toothpicks in as "bee sticks" dry (for example, 1 bottom from a Styrofoam cup can be labeled and used by each lab section)

Activity C: Pollination

No additional materials are required.

Activity D: Germinating the F_2 Seeds

1 Week Before Lab Topic 9 Is Scheduled

Materials for each student team (not provided in Fast Plants kit)

- 1 seed collecting pan (half of a Petri dish can be used)
- 1 seed envelope (to store extra seeds not immediately sown for germination)
- 1 Petri dish
- 1 piece of filter paper to fit Petri dish
- 1 water reservoir: May be shared by 3 teams. The Petri dishes will be placed on end in the water reservoir so that the filter paper can be kept moist. The Growing Instructions booklet suggests using the bottom of a 2-L soft drink bottle.
- 1 wash bottle of water (may be shared by teams)

Activity E: Collecting and Interpreting Data from the F_2 Generation

The Week Lab Topic 9 Is Scheduled

Materials for each student team (not provided in Fast Plants kit)

- Forceps

Reordering kits from Carolina Biological Supply

The watering system and quads can be reused. Rinse everything thoroughly. Use detergent or 10% chlorine bleach (to eliminate algal or fungal growth) only if necessary, and then rinse until no traces remain. Residual detergents or bleach will be harmful to the plants.

Refill kits that include seed, pollination materials, labels, and fertilizer can be ordered from Carolina Biological Supply (#15-8775).

Options for Replacing Materials in Kits

It is possible to set up the Fast Plants experiment without purchasing materials (except seeds) from Carolina Biological Supply. After the first time you prepare for this lab topic, you may want to reduce expenses by trying the following alternatives. These suggestions are adapted from *Preparation Guide for Investigating Biology* by Judith Morgan and Eloise Carter (Benjamin/Cummings, 1993). Each item that comes in the kits is also available separately from Carolina Biological Supply, so items that are used up (e.g., fertilizer) or damaged can be replaced individually.

Seeds

Seeds are available only from Carolina Biological Supply. The experiment described in Lab Topic 9 requires F_1 anthocyaninless yellow-green (#15-8890 and 15-8891).

Watering

The watering trays are rectangular plastic boxes with lids. One end of the lid is cut out to allow the watering mat to dip into the reservoir. Plastic containers made for food storage could be used.

The mats and wicks can be made from Pellon, which is available at fabric stores. Pellon contains sizing and should be washed thoroughly before use. The mat should be large enough to fit the top of the tray and extend down into the reservoir. Refer to Figure 9.5 in the Lab Investigations manual to see the shape of the wicks.

Copper sulfate

To replace the copper sulfate squares that prevent algal growth, add 4–5 drops of 20% copper sulfate solution to the water in each tray. To make the solution, dissolve 20 g $CuSO_4$ in 100 mL water.

Lights

There is no substitute for the 6-tube light bank to provide the required lighting, but fortunately this is a one-time expense. It can be less expensive to make your own than to buy the ones sold by Carolina Biological Supply. The rack can be constructed from wood or from lightweight PVC pipe. The light stands sold by Carolina Biological Supply are made of PVC pipe.

Quads

The Styrofoam quads are available separately from Carolina Biological Supply (#15-8960), but the plants can be grown in any small container. For example, the plastic canisters that 35mm film comes in can be used. Make a hole for the wick in the bottom of each canister with a hot probe. Hold 4 canisters together with a rubber band to make a quad. Canisters can be obtained from film developing companies, which usually discard large numbers of them.

Potting soil

Potting soil can be ordered as a separate item from Carolina Biological Supply (#15-8965 and 15-8966). If you buy potting soil locally, it should not consist primarily of peat moss. It should be a fine-textured artificial mix that is available from garden and greenhouse suppliers. If the soil contains large pieces of bark or peat, sift it through a screen. The quads are very small and can't accommodate bark chunks.

Do not use potting soil that has been formulated for African violets.

Do not use soil that has been dug up from a garden.

Bees

Dried bees can be purchased separately from Carolina Biological Supply (#15-8985). You may be able to find a local beekeeper who will donate dead bees.

Reference Material

You might want to purchase the Wisconsin Fast Plants Manual from Carolina Biological Supply (#15-8950). It contains background information on *Brassica* as well as instructions for numerous exercises using different Fast Plants mutants.

Summary of Materials

	Per team of students	Per room or lab section
Disposables		
copper sulfate square	1	
dead bee	1	
fertilizer pellets	3	
filter paper	1	
pot label	1	
potting mix	to fill quad	
seed envelope	1	
seeds (F_1 Fast Plants)	12	
toothpick	1	
Equipment		
light bank		1
Labware		
forceps	1	
glue	1 tube (share)	
mat for watering tray	1 per 8 teams	
marker (waterproof)	1	
Petri dish	1	
quad	1	
seed collecting pan (can use Petri dish half)	1	
Styrofoam piece		1
plastic pipet	1	
wash bottle	1	
water reservoir (for germinating seeds)	1 per 3 teams	
watering tray	1 per 8 teams	
wicks	4	

Microorganisms and Disease

It is recommended that students work in pairs for this lab topic, but check with the instructor to determine team size and the number of teams per lab section.

Exercise 10.1: General Features of Bacteria

Activity A: Bacterial Shapes

Materials for each student team

- 1 each: Prepared slides of coccus, bacillus, and spirillum bacteria (e.g., Carolina Biological Supply #Ba 048A, Ba 038A, and Ba 028) *or* a single slide showing all three bacterial types (e.g., Carolina Biological Supply #Ba 029). Carolina Biological Supply and other companies offer numerous slides of bacteria that have been stained by different techniques. If your department already has slides that illustrate the morphological types of bacteria, there is no need to purchase the specific slides suggested here.
- Prepared slide of bacterial capsule (e.g., Carolina Biological Supply #Ba 016A)
- Compound light microscope: Microscopes that have oil immersion lenses are preferable for studying bacteria. However, students can see what they need to see at high power (for example, 430×). Our student microscopes don't have oil immersion lenses, so we put one microscope with an oil lens in each lab and have the instructor set up a demonstration slide.
- Lens paper
- Immersion oil (if students have microscopes with oil immersion lenses)

Activity B: Colonies

Materials for each lab section

- 12 Petri dishes of nutrient agar showing bacterial growth: At least 1 week before lab, make up the nutrient agar according to the directions on the jar (generally 25–35 g/L; 1 L makes ~50 plates) and inoculate or expose each plate to a likely source of bacteria (e.g., money, keys, dog paw, drinking fountain, toothbrush, sole of shoe, lips and nose, doorknob, telephone mouthpiece). After inoculation, seal each plate with a strip of Parafilm and store the plates upside down. At room temperature it takes a maximum of 1 week for adequate development. Stop growth while the colonies are still separate from each other by refrigerating the plates. It is important not to let the plates get overgrown (colonies running into each other). The lab exercise requires students to be able to differentiate single colonies. Each lab section should have approximately 12 plates to look at (exposed to 5 or 6 different sources), but make extras in case some don't turn out well or get overgrown. You may need more if the lab sections are large (30–35 students). The plates can be reused by lab sections throughout the week, especially if they are refrigerated each night. The Lab Investigations manual tells students that they should not open the plates, but you should also put up a sign in the lab where the plates are located. To dispose of the plates, autoclave for 15 minutes at 121°C.

- Dissecting microscopes (if available)

Exercise 10.2: Investigating Microorganisms and Disease

Activity A: Gram Stain

The staining stations may be set up for each team of students or placed at several locations around the room for students to share.

Materials for each student team

- Wax marker
- New slide: For many exercises, we use slides that have been used before and washed. However, it is best to use new slides for the Gram stain because it is difficult for students to see the bacteria, and dirty slides make it more difficult. If you must put out used slides, have each student wash the slide with alcohol before use (add a wash bottle of ethanol and a box of cleaning tissues to the materials list).
- Container for waste stain
- Dropping bottle of water, 1 mL per team
- Dropping bottle of crystal violet stain, 1 mL per team (see Recipes section)
- Dropping bottle of Gram's iodine, 1 mL per team (see Recipes section; put a caution label on the bottle)
- Dropping bottle of 95% ethanol, 1 mL per team
- Dropping bottle of safranin stain, 1 mL per team (see Recipes section)
- Toothpick: Use the flattened rather than the round kind so students can use the broad end to obtain and spread tartar on a slide.
- Alcohol lamp: Students need a small flame to heat-fix their slides. Alcohol lamps are safer and easier to use than Bunsen burners.
- Clothespin: Get the spring clip type. Students will use it to hold a slide while passing it through the flame of an alcohol lamp.
- Book of matches
- Ashtray: Students need a safe place to put used matches after lighting the alcohol lamp. We use aluminum weigh boats.
- Small Petri dish
- Large Petri dish: The small Petri dish (60 × 15 mm) is set inside the large Petri dish (150 × 20 mm). The student will place a slide on the small dish and flood it with stain. The large dish catches the excess stain.
- Wash bottle of water
- Waste container for stains

Activity B: Antibiotic Sensitivity Test

Materials for all lab sections

The Antibiotic Sensitivity Kit (Carolina Biological Supply #15-4740, not in the index, but appears on p. 56 of the 1994 catalog) provides all the materials needed to set up this demonstration. There are enough materials in one kit to supply a multisection course that meets in several lab rooms. Once the results have been developed, the dishes will last throughout the week unless they are contaminated.

Activity C: Transmission

Materials for each student

- Disposable (pasteur) pipet with rubber bulb: Use the 9" size of pasteur pipets, which reach to the bottoms of the test tubes. If the instructors rinse the pipets thoroughly, they can be reused throughout all the lab sessions. (If you are reusing pipets, put a large beaker by the sink to stand the pipets in to drain.)
- Test tube: Test tubes can also be rinsed and reused, but they should *not* be washed with soap because its alkaline pH will affect the lab exercise. Put a test tube rack by the sink.

Materials for the class

- 10 dropper bottles of phenol red, ~1 mL/student (students can share; see the Recipes section)
- Bottles labeled with numbers: Have as many bottles as there are students in the largest lab section. Except for one, each bottle should contain dilute HCl (see Recipes section). The remaining bottle should contain 0.1N NaOH. Label the bottles with numbers 1–25 (or enough for each student in the lab section to have a bottle). The lab instructors will need to know which bottle contains NaOH. Each student will use ~5 mL of solution.
- Test tube racks by sink

Recipes

Crystal Violet Stain

Each team will use ~1 mL.

To make 1 L:

Solution A

Dissolve 20 g crystal violet in 200 mL ethyl alcohol.

Solution B

Dissolve 8 g ammonium oxalate in 800 mL distilled water.

Mix Solutions A and B.

Gram's Iodine

Each team will use ~1 mL.

To make 1 L:

Dissolve 6.7 g KI (potassium iodide) in 1000 mL distilled water.

Add 3.3 g I_2 (iodine) and stir until dissolved.

Store in brown bottle. Use brown dropping bottles for lab. Put caution labels on the bottles.

HCl for Exercise 10.2, Activity C

Each student will use ~5 mL.

To make 1 liter:

Add 10 mL 0.1N HCl to 990 mL tap water.

Phenol Red Stock Solution

Each student will use ~1 mL of *diluted* phenol red.

To make 1 L of stock solution:

Dissolve 0.4 g phenol red in 10 mL 0.1*N* NaOH. Dilute up to 1000 mL with tap water.

For the transmission exercise, dilute stock solution 1:10 with tap water.

Safranin Stain

Each team will use ~1 mL.

To make 1 L:

Dissolve 2.5 g safranin O in 100 mL 95% ethanol.

Bring volume up to 1000 mL with distilled water.

Infected Person	Contacts		
	First	Second	Third

Figure 10.1
Transparency master for Table 10.4 in Lab Investigations manual.

Summary of Materials

	Per team of students	Per room or lab section
Disposables		
book of matches		1
immersion oil (optional)	1	
lens paper	1 packet	
pasteur pipet with rubber bulb	1 per student	
toothpick	1	
wax marker	1	
Equipment		
alcohol lamp	1	
compound microscope	1	
dissecting microscope (optional)	1	
Labware		
ashtray	1	
bottle (100–500 mL)	1 per student	
clothespin	1	
dropping bottle (brown)	1	
dropping bottles (clear)	4	12
Petri dish (60 × 15 mm)	1	
Petri dishes (100 × 20 mm)		12
Petri dish (150 × 15 mm)	1	
slide (new)	1	
test tube	1 per student	
test tube racks		1–2
wash bottle	1	
waste stain container	1	
Prepared Slides		
slide of bacteria (cocci)	1	
slide of bacteria (bacilli)	1	
slide of bacteria (spirilli)	1	
slide of bacteria (capsule)	1	

Summary of Materials

	Per team of students	Per room or lab section
Solutions/Chemicals		
crystal violet stain	1 mL	
Gram's iodine	1 mL	
HCl	5 mL	
nutrient agar		250 mL
phenol red (diluted)	1 mL	
safranin O stain	1 mL	
NaOH (0.1N)		5 mL
Other		
Antibiotic Sensitivity Kit		1

LAB TOPIC 11

Plant Diversity

There are two types of displays to set up for this lab topic. Each of the four life cycle displays (moss, fern, pine, and angiosperm) should be set up as an individual unit. Students will start at the beginning of a life cycle and follow it through. A minimum of one display is necessary for each life cycle, but two is better if there is space available. If there are more than 24 students in a lab section, two displays of each life cycle will be necessary.

This lab topic includes a complete description of each specimen to be used and diagrams illustrating how to arrange the displays. The materials list for each life cycle tells what kind of microscope (compound or dissecting), if any, is needed for the specimen. If you don't have enough microscopes for all the displays, consult with the instructor. Some specimens can be consolidated onto the same scope or viewed without a scope. Each specimen is also labeled with an explanatory card taped to the microscope or table.

If the slides and specimens listed are not among those currently being used for this course, check with people who teach or prep botany courses at your institution. Many of these materials may already be available.

Use labeling tape to indicate the events that take place between specimens; labels for the tape are indicated on the diagrams (Figures 11.1–11.4) at the end of this lab topic. (Our prep staff refers to this as the "tape lab.") The numbers that are used in the descriptions and diagrams in the Prep Guide to refer to the specimens (M1, M2, etc.) are *not* used in the Lab Investigations manual, so they should not be included in the displays. You may also find it helpful to refer to Figures 11.4–11.7 in the manual for graphic representations of the life cycles.

The diversity displays show representatives of the different plant groups. When possible, living plants should be used, but herbarium specimens may be substituted if necessary. These plants can be displayed with the related life cycle or distributed around the room. Each plant should be labeled with its genus name, common name (if any), and the common name of its division.

Exercise 11.1: The Alternation of Generations Life Cycle

No materials are needed for this exercise.

Exercise 11.2: Seedless Nonvascular Plants

Materials for the entire class

Diversity Display

- Liverwort under dissecting scope (several specimens, depending upon size of class): If you do not have liverworts available from your greenhouse or by collecting locally, they can be ordered from Carolina Biological Supply. Order *Marchantia* (#15-6540) or *Riccia* (#15-6580).

Specimens can be kept for several months if they are stored in a cool, moist place. We often store liverworts in plastic bags in the refrigerator. Supply them to the labs in a covered dish to preserve moisture. Students will remove the cover to view the specimen.

Moss Life Cycle Display (see Figure 11.1)

- **M1**: Small Petri dish containing moss plant on dissecting scope labeled "moss gametophyte." Use any moss plant (the green, leafy part). If you don't have any available locally, order *Polytrichum* from Carolina Biological Supply (#15-6730). Separate out a single plant and put it in a small Petri dish to keep it moist. If it has a sporophyte attached (see M4), remove it. Like liverworts, mosses can be kept for a long time under cool, moist conditions.

- **M2**: Prepared slide of moss antheridia (Carolina Biological Supply #B361) on compound scope labeled "male gametangium."

- **M3**: Prepared slide of moss archegonia (Carolina Biological Supply #B363) on compound scope labeled "female gametangium."

- **M4**: Small Petri dish containing moss sporophyte on dissecting scope labeled "moss sporophyte." The sporophyte is the thin brown (sometimes red or green) structure arising from the top of some moss plants. Separate the sporophyte from the rest of the plant and put it in a small Petri dish. You can get moss plants with sporophytes from Carolina Biological Supply (#15-6695). Once you have the sporophytes, they will keep forever.

- **M5**: Slide of moss spores on compound microscope labeled "spores." You can make the slide with moss spores purchased from Carolina Biological Supply (#15-6660) or get spores from a moss plant as follows: Find a moss sporophyte and place its top end in the middle of a microscope slide. There is a swelling called a capsule at the tip of the sporophyte (Figure 11.4 in the Lab Investigations manual should help you identify it). Open the capsule by slicing the end off with a razor blade and squeeze it. Check under the dissecting scope to see whether the spores, which look like dust particles, have emerged. If so, remove the capsule and the rest of the sporophyte tissue so only the spores remain on the slide. If you have purchased moss spores, dust the spores onto a slide using a cotton swab. Place a coverslip over the spores and seal it with paraffin, clear fingernail polish, or Permount. A slide prepared in this way is permanent.

- **M6**: Prepared slide of moss protonemata (Carolina Biological Supply #15-6680) on compound scope labeled "germinating spores."

- **M7**: Small Petri dish containing moss plant (same as M1).

Materials for each student

In addition to the displays that are set up for the entire class, each student needs a colored pencil (any color).

Exercise 11.3: Seedless Vascular Plants

Activity A: Club Mosses

Diversity Display

- *Selaginella* (Carolina Biological Supply #15-7015 or a local specimen). *Lycopodium* (Carolina Biological Supply #15-6980 or 15-6990 or locally collected specimens) can also be placed with this display.

Activity B: Horsetails

- *Equisetum* (horsetail) specimen: If you cannot collect this locally, you should purchase the Fern Allies set from Carolina Biological Supply (#15-6950); Carolina Biological Supply only sells horsetails as part of this set, which also includes *Selaginella* and *Lycopodium*.

Activity C: Ferns

- Small aquatic ferns (*Azolla, Salvinia, Marsilea*), if available, can be used for a diversity display.

Fern Life Cycle Display (see Figure 11.2)

- **F1**: Fern plant labeled "fern sporophyte." You can use a fern plant of any variety (though preferably not aquatic). If you can find a plant with sori on the undersides of the leaves, use it (sori look like brown dots arranged in rows).
- **F2**: Slide of fern spores on compound microscope labeled "fern spores." The slide may be purchased from Carolina Biological Supply (#15-6860) or collected from a fern plant as follows: Find a sorus on the underside of a leaflet and scrape a little of it into the well of a concavity slide. Under the microscope you will be able to see the sporangia (see Figure 11.5 in the Investigative Lab manual) with dustlike spores coming out of them. Leave the sporangia on the slide. Seal a coverslip over the concavity with Permount, paraffin, or clear fingernail polish. If you purchase spores you can use a flat slide.
- **F3**: Preserved fern prothallium (Carolina Biological Supply #15-6876) in 50% ethanol in a small Petri dish on dissecting scope labeled "fern gametophyte." After the lab session, return the specimen to its bottle; it can be used in future semesters.
- **F4**: Prepared slide of fern prothallium (Carolina Biological Supply #B 410). This is a whole mount prepared slide of a fern prothallium showing the egg- and sperm-producing structures. Put it on same dissecting scope as F3.
- **F5**: Preserved fern gametophyte with young sporophyte in 50% ethanol in a small Petri dish (*or* prepared slide of same) on dissecting scope labeled "gametophyte with young sporophyte" (Carolina Biological Supply #15-6880). A prepared slide (Carolina Biological Supply #B 415) may be substituted for the preserved specimen. If you use the preserved specimen, return it to its bottle after the lab sessions so it can be used in future semesters.
- **F6**: Fern plant labeled "fern sporophyte." This is another fern plant like F1.

Exercise 11.4: Vascular Seed Plants—Gymnosperms

Activity A: Cycads

Diversity Display

- *Zamia* plant (Carolina Biological Supply #15-7145 or 15-7146)

Activity B: Gingko

Diversity Display

Gingko trees are commonly planted in cities and on college campuses, so you may be able to find a living specimen (just use a small branch), or you can use a herbarium specimen. If you are collecting in the fall, avoid stepping on or touching the fleshy seed coats produced by female trees. (They aren't dangerous; they just have a powerful and lingering odor.)

Activity C: Conifer

Pine Life Cycle Display (see Figure 11.3)

- **P1**: Small branch of any species of pine tree labeled "pine sporophyte." Put it in a flask of water to keep it fresh. (Alternatively, a herbarium specimen can be used.)
- **P2**: Male pine cones labeled "male cones." The male cones are much smaller than female cones and are softer. They shed massive amounts of pollen in the spring. The female cones are larger and woody; they are what most people think of as typical pine cones.

- **P3**: Prepared slide of a pine staminate (male) cone in longitudinal section (l.s.; Carolina Biological Supply #B 495) on dissecting scope labeled "section through male cone showing sporangium."
- **P4**: Slide of pine pollen on compound microscope labeled "male gametophytes (pollen)" (Carolina Biological Supply #B 503 or B 503A). Alternatively, you can make your own slide: Put a little pollen from a male cone onto a slide, put a coverslip over it, and seal the edges with paraffin, nail polish, or Permount.
- **P5**: Female pine cones labeled "female cones."
- **P6**: Slide of pine ovulate cone (l.s.) on dissecting scope labeled "section through female cone showing sporangium" (Carolina Biological Supply #B 500, B 504A, or B 504B).
- **P7**: Slide of pine archegonium (Carolina Biological Supply #B 506 or B 506C) on compound microscope labeled "female gametophyte."
- **P8**: Pine seeds labeled "seeds" (if available). The seeds develop inside the "leaves" of the female cone and resemble the "leaf" in size and shape but are flatter. They separate from the cone when they are mature and can be collected in the fall.
- **P9**: Small branch of pine tree labeled "pine sporophyte" (same as P1).

Exercise 11.5: Vascular Seed Plants—Angiosperms

Angiosperm Life Cycle Display (see Figure 11.4)

- **A1**: Plant with flowers labeled "angiosperm sporophyte," any flowering plant (though it does not need to be in flower). A lily is nice to use because the microscope slides are from lily and the ovaries are large (for A5).
- **A2**: Prepared slide of lily anther tetrads in cross section (Carolina Biological Supply #B 685) on compound scope labeled "spores produced by meiosis inside anther."
- **A3**: Prepared slide of lily anther containing pollen grains (Carolina Biological Supply #B 686) on compound scope labeled "male gametophytes (pollen) develop from spores."
- **A4**: Slide of fresh anther with pollen on dissecting scope labeled "mature anther with pollen." The anther can come from any flower. You should be able to see the pollen grains on the outside of the anther. If the anther is small, put it in the well of a concavity slide and seal a coverslip over it. If it is too large for that, put it in a small Petri dish.
- **A5**: Fresh ovary cut transversely labeled "ovules inside ovary." Any species can be used, but larger is better. The ovary is located in the center of the flower either above or below the petals. It is green and is usually vase-shaped. Cut the ovary across its center. You should see the white ovules inside. Put it in a small Petri dish.
- **A6**: Prepared slide of lily ovulary (Carolina Biological Supply #B 701) on compound scope labeled "spores produced by meiosis inside ovary."
- **A7**: Prepared slide of mature lily ovary (Carolina Biological Supply #B 709C) on compound scope labeled "female gametophyte."
- **A8**: Prepared slide of lily pollen tubes (Carolina Biological Supply #B 691) on compound microscope labeled "pollen tubes bring sperm nuclei for double fertilization."
- **A9**: Small Petri dish of seeds (from any angiosperm) labeled "seeds."
- **A10**: Plant labeled "angiosperm sporophyte" (same as A1).

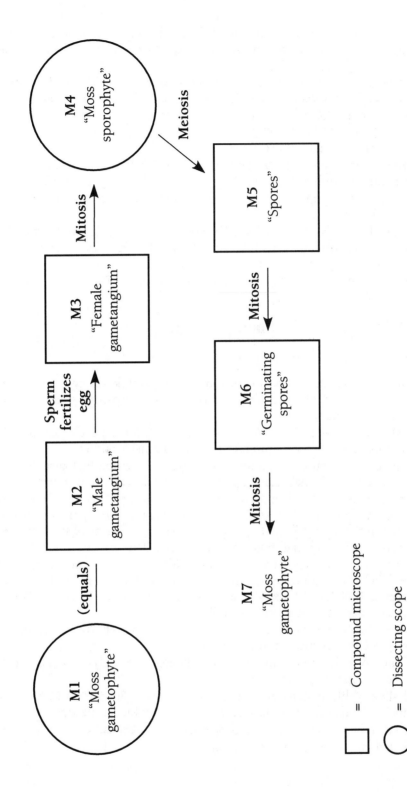

Figure 11.1
Moss life cycle display.

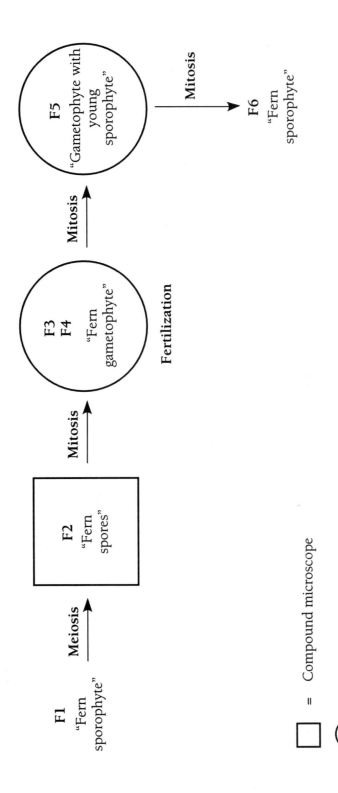

Figure 11.2
Fern life cycle display.

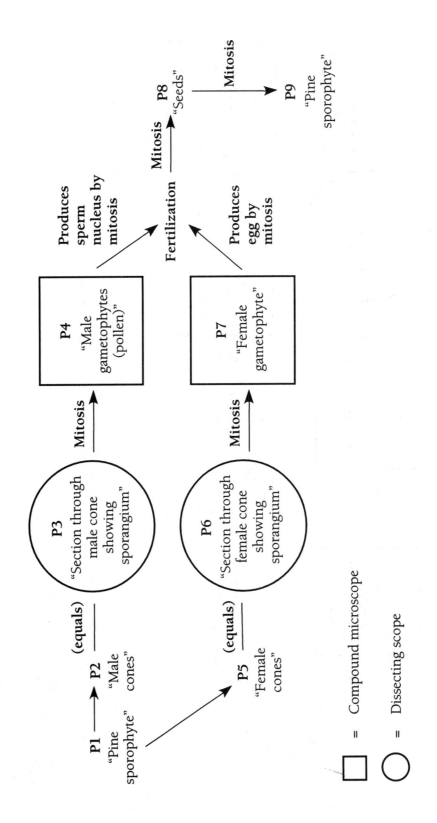

Figure 11.3
Pine life cycle display.

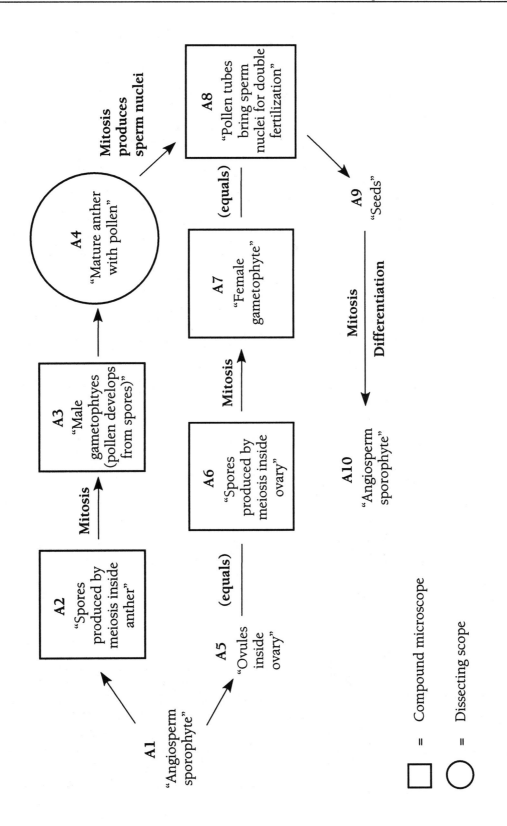

Figure 11.4
Angiosperm life cycle display.

Summary of Materials

	Per student	Per room or lab section
Equipment		
compound microscopes		10
dissecting microscopes		8
Labware		
colored pencil	1	
Living Specimens		
angiosperm (flowering plant) plants		2
angiosperm (flowering plant) seeds		several
anther from flower		1
fern plants		2
fern spores		many
Gingko		1
horsetail (*Equisetum*)		1
liverwort		1
moss plants		2
moss spores		many
moss sporophyte		1
ovary from flower		1
pine branches		2
pine cones (female)		a few
pine cones (male)		a few
pine seeds		several
Selaginella (club moss)		1
Zamia (cycad)		1

✱ NOTE: This list shows the materials needed if there is one of each of the life cycle displays. If you plan to use more displays, increase the quantities of materials.

Summary of Materials (Continued)

	Per student	Per room or lab section
Prepared Slides		
fern prothallium		1
lily anthers (#B 685)		1
lily anthers (#B 686)		1
lily ovary		1
lily ovulary		1
lily pollen tubes		1
moss antheridium		1
moss archegonium		1
moss protonemata		1
pine archegonium		1
pine cone l.s. (female)		1
pine cone l.s. (male)		1
pine pollen		1
Preserved Specimens		
fern gametophyte with young sporophyte		1
fern prothallium		1

LAB TOPIC 12

Animal Diversity I

This lab topic includes exercises in which students perform some procedure using an animal that represents its phylum. In addition, there should be specimens on display in the lab. This lab topic describes the specimens that are named specifically in the Lab Investigations manual for each exercise. The instructor may want to include other animals in the displays to illustrate greater diversity. You may already have many animal specimens in stock that can be used, or you may be able to borrow specimens from other courses. Consult with the instructor about which animals are appropriate for this lab topic. Labels should identify the phylum and common name of each animal.

Since students will not need to spend very long with each specimen, the instructor may want you to have only one or two of each. If your lab sections are very large (25–35 students), additional specimens will be needed. To avoid traffic jams, it will probably be better to place small groups of animals in convenient locations rather than make one big display.

If at all possible, use living animals instead of preserved ones. A saltwater aquarium with half a dozen invertebrates is much more effective in promoting student interest and appreciation for animal diversity than any collection of preserved animals.

Students will work in teams of 2–4 for these exercises. Check with the instructor about the number of teams per lab section.

Exercise 12.1: Animal Characteristics

No materials are necessary for this exercise.

Exercise 12.2: Phylum Porifera—Sponges

Materials for each student team

- Small piece of preserved specimen of *Leucosolenia* (Carolina Biological Supply #P 26): A single sponge should be enough for 8–12 teams of students.
- Small piece of preserved specimen of *Spongilla* (Carolina Biological Supply # P 42): A single sponge should be enough for 8–12 groups of students.
- Forceps
- 2 microscope slides
- Dropping bottle of 0.1N HCl: Put a caution label on the HCl to warn students that it is a strong acid.
- Prepared slide of a sponge (e.g., *Grantia*, Carolina Biological Supply #Z 520): Teams can share these, so it isn't necessary to have one per team.
- Compound microscope

Exercise 12.3: Phylum Cnidaria—Jellyfish, Sea Anemones, Corals

Materials for each student team

- Living hydras: Available from Carolina Biological Supply (#L 55 is the least expensive, but any species can be used). Order the hydras to arrive several days before the lab topic and do not feed them. They can be kept at least 2 weeks without feeding. Students will attempt to feed them during the lab, so they must be hungry. Supply them to the labs in a container labeled "hungry." Alternatively, you can minimize student handling by putting the hydras in watch glasses for each team of students.
- Container marked "fed": Students will return hydras to this container after feeding. They should not be reused immediately, but hydras used in Monday or Tuesday labs may be willing to eat later in the week. One container will serve the entire class.
- Small watch glass
- Dropper (medicine dropper or the plastic droppers that come from Carolina Biological Supply with the animals)
- *Daphnia* (Carolina Biological Supply #L 565) or brine shrimp (Carolina Biological Supply #L 608): To feed the hydras. If you want to hatch your own brine shrimp (eggs are available at pet supply stores as well as through Carolina Biological Supply), allow 2 days for hatching.
- Dissecting scope
- 1–2 preserved sea anemones (e.g., Carolina Biological Supply #P 160C): Label each specimen with its phylum and common name.
- 1–2 preserved *Obelia* (e.g., Carolina # P 92C and P 93F represent the two phases of the life cycle): Label each specimen with its phylum and common name.

Exercise 12.4: Phylum Platyhelminthes—Flatworms

Materials for each student team

- Planarian: Brown planaria (Carolina Biological Supply #L 215) are relatively inexpensive and can be used for this exercise, but white planaria (Carolina Biological Supply #L 225) are superior for demonstrating the digestive tract. Check with the instructor as to whether this is worth the added expense for your course. Do not use black planaria. As with the hydras, students will be trying to observe feeding behavior, so the animals should not be fed after they arrive. They can last up to 3 weeks without food. You may supply the planaria to the labs in a large container or give each student group its own watch glass with a planarian in it.
- "Fed" container: Planaria that have been fed by students early in the week can be reused for labs later in the week.
- Watch glass
- Tiny bit of hard-boiled egg yolk (colored red) or liver: Either can be used as planaria food, but keep in mind that liver will spoil very quickly. Either should be chopped very fine. If you use egg yolk, mix in a little red food coloring. This helps students see the food in the digestive tract of the planaria. The Annotated Instructor's Edition also mentions that planarians will ingest human blood. If the instructor decides to make the sacrifice, he or she will need a sterile lancet and an alcohol swab. Be sure you put these supplies where only the instructor has access to them (or give them directly to the instructor). Because of the danger of blood-borne pathogens, students should not be permitted to use their own blood. Provide a container for instructors to use for proper disposal of lancets.
- Forceps
- Dissecting microscope
- 2 preserved tapeworms (e.g., Carolina Biological Supply #P 230C and P 245C): Label the specimens with their phylum and common names.

Exercise 12.5: Phylum Nematoda—Roundworms

Materials for each student team

- 1 drop of vinegar eel culture (e.g., Carolina Biological Supply #L 258): The culture can be supplied to students in the container it arrives in or divided into smaller containers. If you do the latter, keep in mind that most of the vinegar eels will be found on the bottom of the container.
- Microscope slide and coverslip
- Compound microscope
- 1–2 preserved specimens of a parasitic nematode (e.g., Carolina # P 265C, P 275C, P 280C, or P 291): Label each specimen with its phylum and common name.

Exercise 12.6: Phylum Mollusca—Mollusks

Materials for the entire lab section

- 2 or 3 fresh clams or mussels: Students will use these to identify internal structures. Many large grocery stores sell clams or mussels in the seafood department. If you are not able to get fresh specimens, check with the instructor about whether to buy preserved specimens. Consult with the instructor about how to supply the clams to lab teams. You may be expected to peg the clams ahead of time (that is, insert a wooden peg once the clam is partially opened) or actually open them before they are placed in the labs. Do this over a sink or in a dissecting pan; a lot of liquid will come out as you open the clam. To open a clam, use a knife or strong scalpel to cut away as much as you can of the hinge that holds the two shells together. You should be able to insert the blade into one end of the hinge and make an opening big enough to force in a screwdriver. Insert the screwdriver; then twist it to pry the shells apart slightly. Either leave that screwdriver in place or insert some kind of wedge to hold the shell open. Then insert a screwdriver on the other side of the shell and twist it to open the shell further. If the instructor only wants you to peg the clam, you can now insert a small wooden peg to hold the shells apart. At this point you should be able to slide a scalpel between the shells and cut the two muscles holding it together. Next, open the shell slowly until you can see in. The mantle will be clinging to the top shell. Carefully run the scalpel underneath it to free it from the shell. You will have to do a little extra cutting where it is attached to the muscles. Finally, bend the hinge back so that the clam can be laid down flat for display.
- Dissecting pan with cover for each clam
- Blunt probe with each clam (e.g., Carolina Biological Supply #62-7400)

Exercise 12.7: Phylum Annelida—Segmented Worms

Materials for each student team

- Living earthworm (e.g., Carolina Biological Supply #L 408): Small or medium-sized worms are fine since they will not be dissected. Before ordering, you might try to find worms locally at a bait shop or pet store.
- 2 preserved *Nereis* (clam worm; e.g., Carolina Biological Supply #P 412C or P 413C): Label the specimens with their phylum and common names.

Display of "Unknown" Animals

The instructor may wish to have a display of unlabeled specimens for students to identify. The purpose of the display is to have students use the characteristics they have observed in the animals presented in the lab session in order to determine the phylum of an animal that has not been classified for them. The unknowns should therefore be similar, but not identical, to the animals displayed for each exercise. Here are some suggestions for animals to use (catalog numbers from Carolina Biological Supply).

Phylum Porifera (Sponges)

Grantia (#P 20)

Rhabdodermella (#P 24)

Halichondria (#P 39)

Any of the commercial sponges (#P 45 through P 52)

Phylum Cnidaria (Jellyfish, Sea Anemones, Corals)

Hydrozoan collection (#P 185)

Medusae collection (#P 190F)

Anemone-coral collection (#P 195C)

Any of the dry corals (#P 173, P 175A, P 179, or P 180A)

Phylum Platyhelminthes (Flatworms)

Bdelloura (#P 215)

Sheep liver fluke (#P 225C)

Tapeworm (#P 230C or P 245C, whichever wasn't already used in the flatworm display)

Phylum Nematoda (Roundworms)

Ascaris (#P 265C), *Macracanthorhynchus* (#P 275C), *Ancyclostomas* (#P 280C), or *Trichinella* (#P 290), whichever haven't already been used in the roundworm display

Phylum Mollusca (Mollusks)

Limpet (#P 470C)

Nudibranch (#P 472C)

Common oyster drill (#P 475AC)

Conch (#P 478C)

Periwinkle (#P 485C)

Land snails (#P 485AC)

Limax (#P 494C)

Razor clam (#P 505C)

Long-necked clam (#P 506C)

Mytilus (#P 509C)

Oyster (#P 510C)

Scallop (#P 512C)

Squid (#P 530S or P 530D)

Phylum Annelida (Segmented Worms)

Leech (#P 425 is the small inexpensive specimen; P 416 is up to 7" long)

Lugworm (#P 434C)

Clymenella (#P 440C)

Summary of Materials

	Per team of students	Per room or lab section
Disposables		
clams or mussels (fresh)		2–3
egg yolk (boiled, dyed red with food coloring) or liver	tiny bit	
Leucosolenia sponge (preserved)	small piece	
Spongilla sponge (preserved)	small piece	
vinegar eel culture	1 drop	
Equipment		
compound microscope	1	
dissecting microscope	1	
Labware		
blunt probes		2–3
dissecting pans		2–3
dropper	1	
dropping bottle	1	
"fed" container (hydras)		1
"fed" container (planaria)	1	1
forceps	3	
microscope slides	2	
watch glass	1	
Living Specimens		
Daphnia or brine shrimp	1	
earthworm	1	
hydra	1	
planarian	1	
Preserved Specimens		
Nereis (clam worms)		2
Obelia		1–2
parasitic nematodes		1–2
sea anemones		1–2
tapeworms		2
unknowns (see discussion on pages 66–67)		
Prepared Slides		
sponges		3–6
Solutions/Chemicals		
0.1N HCl	~1 mL	

Animal Diversity II

✳ NOTE: Advance planning and ordering is required for this lab topic. Read the following prep directions carefully as soon as you know the lab schedule.

Students will work in teams of 2–4 for these exercises. Check with the instructor regarding the number of teams per lab section.

Exercise 13.1: Phylum Arthropoda

Materials for each student team

If possible, use specimens that have been collected locally, killed in a kill jar or freezer, and preserved dry rather than specimens from a biological supply house that have been preserved in a liquid preservative.

Students will observe the external features of these animals, so the same specimens can be used for multiple lab sections. But since many of these specimens are delicate, plan to have extras on hand to replace those that are damaged by student handling.

Except for the grasshopper and crayfish, there are no specific animals mentioned in the Lab Investigations manual, so any members of these groups may be used. It is not necessary to give the same species to each student group.

Each specimen should be labeled with its phylum and common name (or species name if no common name is known).

- Grasshopper: Any species is acceptable, so just get the least expensive kind (Carolina Biological Supply #P 727).
- 2 or 3 other insects: Carolina Biological Supply has a whole variety of insects for sale (see catalog index); consult with the course instructor about choices.
- 1 or 2 arachnids: Arachnids include spiders, mites, ticks, and scorpions. Carolina Biological Supply offers an arachnid collection that contains a horseshoe crab, scorpion, spider, and tick (#P 685); they also have a selection of various spiders and ticks (see catalog index) and scorpions (#P 655 and P 656).
- Crayfish: Carolina Biological Supply #P 595C or P 595D (the smallest size) is adequate. Color-injected crayfish are not needed for this lab.
- 1 or 2 other crustaceans: Crustaceans include crabs, lobsters, crayfish, shrimp and pillbugs (see catalog index); Carolina Biological Supply also offers a crustacean collection (#P 635).

- Centipede (e.g., Carolina Biological Supply #P 643)
- Millipede: Carolina Biological Supply has millipedes in three different sizes. The small size (#P 646) is fine.
- Teasing needle
- Dissecting scope

Exercise 13.2: Phylum Echinodermata

Materials for each student team

Students will observe the external features of these animals, so the same specimens can be used for multiple lab sections.

- Sea star (starfish; Carolina Biological Supply #P 335C, 335D, or 335S)
- Brittle star (Carolina Biological Supply #P 351)
- Sand dollar or sea urchin (Carolina Biological Supply #P 377 or #P 359)
- Sea cucumber (Carolina Biological Supply #P 380C)

Carolina Biological Supply also offers an echinoderm collection that includes the specimens needed (#P 385). It is not necessary for all groups to have the same species, so if you already have a collection of specimens you can use them. Also, the instructor may want to have additional specimens displayed.

If possible, a saltwater aquarium containing echinoderms is an excellent addition to this lab exercise.

Exercise 13.3: Phylum Chordata

Activity A: Jawed Fishes

Materials for each student team

- Dissecting microscope
- Dish containing several living fish eggs: Eggs from the Japanese medaka can be ordered from Carolina Biological Supply (#L 1421A; you must give the shipping date with the order). One vial of 12 eggs is sufficient for a lab room, since you can divide them up among the groups. The amount of embryo rearing medium that comes with the eggs is sufficient if you put the eggs in small Petri dishes for the student groups. Should you need more, directions for preparing the medium are in the booklet that comes with the eggs. It is best to use eggs that are at least 2 days old. Students will be able to see the heart beating at that time, and in later stages they can see blood circulating and the fish moving its pectoral fins. Eggs will begin hatching around the 11th day. Until then, eggs can be used in successive lab sections.

The booklet that comes with the medaka eggs gives detailed instructions for hatching and raising the fish. If you are interested, you can breed them and keep your own egg supply going. Order adult medakas for this, since it takes several months for hatchlings to reach maturity. Medakas and similar fish are widely used for developmental studies. If your department teaches a laboratory course in developmental biology, they may already breed fish and be able to supply you with eggs. Zebra fish, for example, can be used in place of medakas.

Activity B: Amphibians

Materials for each student team

- Dissecting scope
- Dish containing several living frog eggs: Eggs can be obtained from Carolina Biological Supply (#L 1490). One unit has 75–100 eggs, which may be divided up among the teams in a lab

section. You will need spring water to keep them in. Supply the eggs to the labs in culture dishes so students will be able to view them on the dissecting scope. Since students merely observe the eggs, they can be used by successive lab sections.

- Amphibian display (1 display for the lab room): The amphibian display should contain living animals, which can be any representatives of this group. You may be able to obtain frogs, toads, salamanders, or newts. Carolina Biological Supply sells living organisms, and many pet stores also carry different types of frogs. Axolotls (mud puppies) and caecilians (legless amphibians) are aquatic amphibians that can also be found in some pet stores. You might also be able to arrange to borrow animals for display from a pet store. The display should have a sign on it that identifies the animals and clearly states that students may not handle the animals.

Before you purchase animals, be sure you have a suitable habitat prepared for each one, and also have food on hand. Carolina Biological Supply sells complete habitats for various animals (see catalog index). You should either be prepared to maintain the animals or to find homes for them when the lab is finished. Many K–12 teachers keep animals in their classrooms and would be glad for the donation. If you aren't in contact with any teachers, call schools and leave the information for posting. The education department at your institution may have student teachers who may be willing to take animals to their schools. Some of these animals can make good pets; offer them to the children of faculty and staff or to graduate students. They can also be adopted by the biology lab students. Whatever you do, *don't* release animals into the wild unless you have collected them from the same site.

Most colleges and universities have regulations regarding the use of living vertebrates, even if they are used only for display purposes. Typically, an animal-use protocol must be filed and approved, including such information as the source of the animals, how they will be used, how they will be fed and cared for, and the means of disposal. Staff members who are responsible for their care may be required to attend a training session. Disposal of frog tadpoles are also covered under these regulations. Since the regulations are generally aimed at researchers who perform animal experimentation, it is usually not a problem to obtain permission for the animal uses that are described in this lab manual. If you are not familiar with the policies at your institution, check with your department administrator.

Activity C: Reptiles

Materials for the lab room

- Reptile display: The reptile display should contain living animals, which can be any representatives of this group. You may be able to obtain lizards, snakes, and turtles for the reptile display. Carolina Biological Supply sells living organisms, and many pet stores also carry different types of turtles, anole lizards, and snakes. You might also be able to arrange to borrow animals for display from a pet store. The display should have a sign on it that identifies the animals and clearly states whether or not students may handle the animals.

NOTE: Please read the notes under Activity B for a discussion about using living vertebrates.

Activity D: Birds

Materials for each student team

- Dissecting microscope
- Fertile chicken egg in dish of chick Ringer's solution: You may be able to find fertile chicken eggs locally or you can purchase them from Carolina Biological Supply (#L 1726). Check Carolina Biological Supply's ordering instructions when you plan your lab: They require 2 weeks' notice, and ship only 1 day a week. If possible, use eggs that are 2–7 days old. You will need a culture dish and about 150 mL of chick Ringer's solution for each egg (see Recipes section). If the number of eggs is limited, student groups may share them. The embryos will live for several hours in Ringer's, so the same eggs can be used in different lab sections. Keep the eggs in a 37° C incubator until they are used in lab.

The instructor may open the eggs himself or herself at the beginning of class, or the prep staff may be asked to do it. To open an egg, first fill a culture dish at least half full with chick Ringer's. Crack the egg gently on the side of the dish. Hold it under the Ringer's solution to open it, so the egg slides right out into the solution.

You will have to file an animal-use protocol for using chicken eggs as well as for the amphibians and reptiles. When the labs are all finished, destroy any remaining eggs by placing them in the freezer (if you don't they may hatch). Then dispose of them according to the procedures used at your institution.

Activity E: Mammals

No materials are needed.

Recipe

Chick Ringer's Solution

~150 mL used with each fertile egg

To make 1 L:
Dissolve the following salts in distilled water:

9.0 g NaCl

0.4 g KCl

0.24 g $CaCl_2$

0.2 g $NaHCO_3$

Summary of Materials

	Per team of students	Per room or lab section
Equipment		
dissecting microscope	1	
Labware		
culture dish	1	
teasing needle	1	
Living Specimens		
amphibian display		1
chicken egg (fertile)	1	
fish eggs*	3–4	
frog eggs*	3–4	
reptile display		1
Preserved Specimens*		
arachnids	1 or 2	
brittle star	1	
centipede	1	
crayfish	1	
crustaceans	1 or 2	
grasshopper	1	
insects	2 or 3	
millipede	1	
sand dollar or sea urchin	1	
sea cucumber	1	
sea star	1	
Solutions/Chemicals		
chick Ringer's	~150 mL	

*These specimens can be reused in successive lab sections.

LAB TOPIC 14

Digestion

✳ NOTE: All solutions should be at room temperature when they are supplied to the labs.

In Exercises 14.1, 14.2, and 14.3 students learn methods for investigating digestive enzymes. In Exercise 14.4 students design their own experiment using one or more of these methods. In Exercise 14.5 they perform their own experiment. The prep for Exercise 14.4 includes giving each student team a basic kit of materials similar to the materials needed for Exercise 14.1, 14.2, or 14.3. Students will also require a variety of additional materials for their experiments. Consult with the instructor about the additional materials that you should have on hand.

Students will work in teams throughout the exercises. The instructor will specify the number of students per team (usually 2–4) and will tell you how many teams to prepare for per lab section. The quantities of materials needed by each team are the same regardless of team size.

Exercises 14.1, 14.2, and 14.3

Materials for each student team, to be used in all 3 exercises

- Test tube rack
- Wax marker
- ~15-mL pancreatic extract in bottle
- ~22-mL distilled water in bottle
- 5-mL pipet labeled "water"
- 5-mL pipet labeled "pancreatic extract"
- Pi-pump for 5-mL pipet

Exercise 14.1: Digestion of Proteins by Trypsin

Materials for each student team

- 2 test tubes
- 2 pasteur pipets with rubber bulbs
- 10-mL graduated cylinder
- Ice chest with ice (can be shared by class)

In addition, the instructor needs the following to make up the 2% gelatin:

- Stirring/heating plate (if a hot plate that also stirs is not available, give instructors a spoon to stir the gelatin)

74

- Magnetic stir bar
- Gelatin powder: Can be preweighed and supplied to the instructors in plastic bags or paper packets; 4 g will make 200 mL, which should be enough for most labs.
- 2 400-mL beakers: The gelatin will be softened in one beaker while water is being heated in the other.
- 100-mL or 250-mL graduated cylinder

Exercise 14.2: Digestion of Lipids by Lipase

Materials for each student team

- 3 test tubes
- 3 small Parafilm squares: Used to cover test tubes, so cut the standard 2" squares into quarters.
- 2 1-mL pipets
- Pi-pump for 1-mL pipet
- 2 mL 5% bile
- 1.5 mL 1% phenolphthalein in 95% ethanol in dropping bottle
- <1 mL 0.1N NaOH in dropping bottle (put caution label on bottles)
- 6 mL vegetable oil: Any type of vegetable oil will work for this exercise, but lighter-colored ones are preferable.

Exercise 14.3: Digestion of Starch by Amylase

Materials for each student team

- 12 test tubes
- 4 small Parafilm squares: Used to cover test tubes, so cut the standard 2" squares into quarters.
- 5-mL pipet
- 15-mL bottle of 0.5% starch
- 4 pasteur pipets
- 2 mL iodine reagent (I_2KI) in brown dropping bottle (put caution label on bottle)
- <1 mL 0.5% HCl in dropping bottle
- pH test paper (4 strips per team): The instructor *may* request this. It should have a range that includes pH 4 to 6.

Exercise 14.4: Designing an Experiment

No materials are needed for this exercise.

Exercise 14.5: Performing the Experiment and Interpreting the Results

Each student team should receive a basic kit for performing the method in Exercise 14.1, 14.2, or 14.3. The basic kits include some materials students have already used for Exercises 14.1, 14.2, and 14.3. Check with the instructor about the number of teams per lab section that will need each kit (generally, each team uses only one kit). Prepare as many basic kits as will be needed for one lab section and put them in the lab room in boxes (or plastic dishpans or some other container). Students can be expected to clean their glassware and reassemble the kits when they are finished, so kits can be reused through multiple lab sections. It would also be useful to have extra glassware, especially test tubes and pipets, available in the lab.

✳ NOTE: Asterisks (*) denote materials that can be reused from Exercises 14.1, 14.2, and 14.3. Quantities of solutions are in addition to the quantities needed for Exercises 14.1, 14.2, and 14.3. See the Summary of Materials for total quantities for the lab topic.

Basic Kit for Exercise 14.1 (Trypsin Experiment)

- 15 test tubes*
- Test tube rack*
- 10-mL graduated cylinder*
- 3 5-mL pipets with pi-pump
- Ice chest with ice*
- 20 mL pancreatic extract*
- 150 mL 2% gelatin solution*

Basic Kit for Exercise 14.2 (Lipase Experiment)

- 15 test tubes*
- 15 small Parafilm squares
- 50 mL pancreatic extract*
- 15 mL phenolphthalein solution in dropping bottle*
- 10 mL 0.1N NaOH (put caution label on bottles)*
- 15 mL vegetable oil*
- 15 mL 5% bile*
- 3 5-mL pipets with pi-pump

Basic Kit for Exercise 14.3 (Amylase Experiment)

- 15 test tubes*
- 3 1-mL pipets with pi-pump
- 3 5-mL pipets with pi-pump
- 20 mL pancreatic extract
- 50 mL starch solution
- 10 mL iodine reagent (I_2KI)

Optional Materials for Amylase Basic Kit

- ~20 mL Benedict reagent in dropping bottle (Carolina Biological Supply #84-8211 and 84-8213)
- 250-mL or 400- mL beaker to use as water bath
- Hot plate

Additional Materials

Each student team will need additional materials. Consult with the instructor regarding materials and quantities that should be available. Students may bring in some materials themselves.

Materials needed for possible student investigations are given below, along with approximate quantities needed per team (each team will do only one investigation). These items can be placed in a central location in the lab room.

Effect of common proteases on gelatin

Proteases (protein-digesting enzymes) are found in meat tenderizer, pineapple, and kiwi fruit (there may be other sources). One container of meat tenderizing powder (for example, Adolph's) should suffice for many student investigations. Supply both fresh or frozen and canned pineapple, if possible. Students will probably use small amounts of fruit—for example, a couple chunks of canned pineapple or a slice of kiwi fruit. One fresh pineapple should be enough for all the student investigations in a lab course.

Effect of pH on activity of trypsin

Supply 0.1N HCl, 0.5% HCl, 1% NaHCO$_3$, and 0.1N NaOH for students to vary pH. Put caution labels on the bottles of 0.1N HCl and 0.1N NaOH. Also supply alkacid paper for measuring pH. Since this type of investigation produces negative results, check with the instructor about whether it will be an option for students.

Different substrates for fat digestion

Different types of cooking oils can be used (e.g., olive, safflower, sunflower, peanut, corn, canola, liquid margarine, and spray-on substitute). Or students can compare liquids or liquifiable foods such as milk, ice cream, half-and-half, whipping cream, and yogurt. Students will use approximately 10–15 mL of each substance.

Fat emulsifiers

Small amounts of mustard (any variety) or raw egg yolk may be tested.

Temperature

Supply means for varying temperature such as water baths, ice, and hot plate, and a thermometer.

Enzyme or substrate concentration

No additional materials are required.

Effect of pH on amylase activity

pH buffers, approximately 10–15 mL of each pH, are needed. Recipes for pH 2, 4, 5, 6, 7, 8, 10, and 12 buffers are given in Lab Topic 3. Provide a 5-mL pipet for each buffer. Put caution labels on bottles of pH 2 and 12. Also supply alkacid paper.

Recipes

5% Bile

Each team will use ~1 mL. An additional 15 mL is needed for each lipase basic kit.

To make 100 mL:
Dissolve 5 g bile salts in 100 mL distilled water.

Bile salts can be purchased from Sigma (#B 8756).

2% Gelatin

Each team will use ~20 mL. An additional 150 mL is needed for each trypsin basic kit.

The instructor will prepare the gelatin solution at the beginning of lab.

To make 100 mL:
Dissolve 2 g gelatin powder in 100 mL water.

Blend the gelatin powder with a little cold water first to wet it. Bring about half the water to a near boil and add it to the gelatin, stirring to dissolve it. Once the gelatin has dissolved, add the rest of the water, which should be cold, while continuing to stir.

0.5% HCl

Each team will use <1 mL. Additional quantities may be needed for "additional materials."

To make 100 mL:
Add 0.5 mL concentrated HCl to 99.5 mL distilled water.

Supply in dropping bottles.

Iodine Reagent (I_2KI)

Each team will use ~2 mL. An additional 10 mL is needed for each amylase basic kit.

To make 1000 mL:
Dissolve 7.5 g KI (potassium iodide) in 1000 mL distilled water.

Add 1.5 g I_2 (iodine) and stir until dissolved.

Store in brown bottle.

Supply in brown dropping bottles. Put caution labels on the bottles.

Pancreatic Extract

Each team will use ~60 mL, including what is needed for their investigations.

To make 1000 mL:
Dissolve 3 g pancreatin (3× U.S.P.) in 1000 mL 30% ethanol. Adjust the pH to 8 with NaOH.

Pancreatin can be purchased from Sigma (#P 1625).

1% Phenolphthalein

Each team will use ~1.5 mL. An additional 15 mL is needed for each lipase basic kit.

To make 100 mL:
Dissolve 1 g phenolphthalein powder in 100 mL 95% ethanol.

Supply in dropping bottles.

0.1N NaOH

Each team will use ~1 mL. An additional 10 mL is needed for each lipase basic kit.

NaOH (sodium hydroxide) can be purchased as 1N or 0.1N solutions.

Supply in dropping bottles. Put caution labels on the bottles.

0.5% Starch

Each team will use ~15 mL. An additional 50 mL is needed for each trypsin basic kit.

To make 1000 mL:

Dissolve 5 g soluble starch (e.g., Fisher #S516-100) in 1000 mL distilled water.

The solution must be heated in order to dissolve the starch. When it has dissolved, the solution will suddenly look clear rather than milky.

Summary of Materials

	Per pair or team of students	Per room or lab section
Disposables		
alkacid paper (optional)	4 strips	
gelatin		~4 g
crushed ice (to chill 2 test tubes)	1 container	
Parafilm squares	7	
pasteur pipets	4	
Equipment		
ice chest		1
stirring or hot plate		1
Labware		
400-mL beakers		2
dropping bottles (for solutions listed below)	3	
dropping bottle, brown (for I$_2$KI)	1	
10-mL graduated cylinder	1	
100-mL or 250-mL graduated cylinder		1
1-mL pipets	2	
5-mL pipets	3	
pi-pump for 1-mL pipet	1	
pi-pump for 5-mL pipet	1	
stir bar		1
stir rod	1	
test tubes	17	
test tube rack	1	
wax marker	1	
Solutions/Chemicals		
5% bile	1 mL	
iodine (I$_2$KI) reagent	2 mL	
0.5% HCl	<1 mL	
pancreatic extract	15 mL	
1% phenolphthalein	1.5 mL	
0.1N NaOH	1 mL	
0.5% starch	15 mL	
vegetable oil	6 mL	
water (distilled)	25 mL	

Circulation

In this lab topic students observe the anatomy of the mammalian circulatory system (Exercise 15.1) and learn methods for measuring pulse rate and blood pressure (Exercise 15.2). In Exercises 15.3 and 15.4 students are asked to design and perform their own investigations of pulse rate and blood pressure. Instructors may choose to have their students do Alternative Exercises instead of Exercises 15.3 and 15.4. Check with the instructor to see which set of exercises you should prepare for.

Exercise 15.1: Anatomy of the Mammalian Circulatory System

Materials for each student team

- Preserved pig heart (Carolina Biological Supply #P 2050C): Sheep hearts are less expensive but smaller and can be used instead (Carolina Biological Supply #P 2140C). The hearts should be cut longitudinally so that the interiors of the chambers can be seen (see Figure 15.2 in the Lab Investigations manual). Preserved hearts can be kept in Carosafe and reused for years.
- Prepared slide of artery and vein (Carolina Biological Supply #H 7045 or #H 7050): If you already have separate slides showing arteries and veins you can use those, but alert the instructor since the Lab Investigations manual refers to a single slide.
- Compound microscope

Exercise 15.2: Measuring Cardiovascular Function in Humans

Materials for each student team

- Blood pressure kit and stethoscope: A basic blood pressure kit contains an aneroid sphygmomanometer and a cuff (e.g., Ward's #14W5011). A stethoscope is also needed to measure blood pressure with this type of apparatus, and part of the exercise requires the separate use of a stethoscope. Blood pressure kits that include a stethoscope are also available (e.g., Ward's #14W5023). Sargent-Welch sells two grades of kits, the basic model (#S-5596-10) and an economy model (#S-5596-05) but neither kit includes a stethoscope. Carolina Biological Supply offers a set (#69-1090) that does include a stethoscope. Stethoscopes can also be purchased separately from all three supply houses. If you have electronic sphygmomanometers that don't require stethoscopes, you will still need to have stethoscopes for Activity A. A container of alcohol and cotton balls should be provided with each kit so that students can clean the earpieces of the stethoscopes.
- Bottle of isopropyl alcohol
- Cotton balls: Each student will use at least one cotton ball to clean the earpieces of the stethoscope before using it.

Exercise 15.3: Designing an Experiment

No materials are needed for this exercise.

Exercise 15.4: Performing the Experiment and Interpreting the Results

The basic kit for this lab topic consists of the same materials that were provided for Exercise 15.2. There may not be a need for additional materials. Check with the instructor about whether you will be expected to provide anything else.

Summary of Materials

	Per team of students	Per room or lab section
Disposables		
cotton balls	1 per student	
Equipment		
blood pressure kit	1	
compound microscope	1	
stethoscope	1	
Preserved Specimen		
pig heart	1	
Prepared Slide		
artery and vein	1	
Solutions/Chemicals		
alcohol for cleaning earpieces of stethoscope		1 bottle

The Sensory System

Students will be working in pairs to perform these experiments.

Exercise 16.1: Taste and Smell

Activity A: Taste Discrimination

Materials for each lab section

Each student will use ~5 mL of each of the following solutions. Except for magnesium sulfate the solutions are "unknowns," so they should be labeled with numbers instead of the name of the chemical. The instructor needs to have a key to the numbers.

- Bottle of 10% NaCl labeled "Solution 1"
- Bottle of 10% sucrose labeled "Solution 2"
- Bottle of quinine labeled "Solution 3"
- Bottle of tap water labeled "Solution 4"
- Bottle of 5% acetic acid labeled "Solution 5"
- Bottle of 10% magnesium sulfate
- 3 oz. disposable cups (8 *per lab team*)
- Cotton swabs (3 *per student*)
- Wax marker

Activity B: Individual Differences in Tastes

Materials for each student

- Taste papers (1 square of each type for each student): These can be obtained from Carolina Biological Supply in packs of 100 strips each. Control (#17-4000), phenylthiocarbamide (PTC) taste paper (#17-4010), sodium benzoate (#17-4020), and thiourea (#17-4030) papers are used in the exercise.

Activity C: Smell Discrimination and Its Influence on Taste

Materials for each student

- Life Savers candy pieces: Use at least three different flavors; fruit flavors work best. Break up the Life Savers into pieces and put them out in clean Petri dishes. (They disappear rapidly, so we only put out enough for a few lab sections at a time.) Plan on using 1 whole Life Saver per student. There are 11 Life Savers in a roll. Alternatively, if you can find fruit-flavored Life Savers "holes," you save yourself the extra chore of breaking the candies into pieces.

Exercise 16.2: Vision

Activity A: Structural Features of the Eye

Materials for each pair of students

- Sheep eye: Preserved sheep eyes can be obtained from Carolina Biological Supply (#P 2130C or P 2130D). It is often possible to obtain pig or other animal eyes from a slaughterhouse. Put them in 70% alcohol. Fresh eyes are preferable because the cornea and lens are not clouded by preservative, but they will last only a limited length of time (a day or two) in the lab before the alcohol affects the cornea and lens.
- Dissecting pan
- Blunt probe
- Disposable latex gloves (Fisher Scientific #11-394-36A-D)

Activity B: Photoreceptors

Materials for the lab section

- Piece of red transparent filter: Filters with good transmission characteristics can be obtained from Edmund Scientific (medium red #823, catalog number G82,015). The same filters are used in the Alternative Exercises for Lab Topic 8 (Photosynthesis). Size of the filter should be adequate to cover the light source used.
- Piece of green transparent filter: Filters with good transmission characteristics can be obtained from Edmund Scientific (medium green #874, catalog number G82,041). Size of the filter should be adequate to cover the light source used.
- 2 light sources: The light sources can be anything that projects a strong enough beam to transmit through the filters. Slide projectors work well (use an empty slide mount if the projector doesn't operate without a slide in it). Strong flashlights or some types of external microscope illuminators can also be used.
- Sheet of white paper (used as a background for the projected beams)
- Color blindness test charts (optional): The Ishihara color blindness test charts can be obtained from Fisher-EMD (#S45551); Carolina Biological Supply offers a similar booklet (#69-4625). However, these tests are not required for the lab exercises and are relatively expensive ($80–$90).

Activity C: The Blind Spot

No materials are needed.

Activity D: Accommodation

Materials for each pair of students

- 30-cm (12") ruler (should have metric markings)

Exercise 16.3: Skin Senses

Activity A: Sensitivity to Pressure

Materials for each pair of students

- Compass: Inexpensive compasses (e.g., Carolina Biological Supply #64-4375) can be used. Insert a teasing needle into the pencil side so that the two points are similar.
- Ruler: Marked in millimeters. Can use the same ruler as used for Exercise 16.2, Activity D.

Activity B: Perception of Temperature

Materials for each pair of students

- 3 1-L beakers
- Thermometer
- Ice bucket
- Water bath (optional)

Students will fill the beakers with water that is 20°C, 33°C, and 44°C. Ice and a water bath (or some other means of heating water) will be needed if students will not be able to get tap water at these temperatures.

Recipes

 SAFETY NOTE: Students will be tasting the following solutions. Be sure to use labware that has never been contaminated by toxic substances throughout preparation of the solutions. The solutions should be supplied to the labs in clean glassware as well.

5% Acetic Acid

Each student will use ~5 mL.

To make 1 L:

Add 50 mL glacial acetic acid to 950 mL tap water.

10% MgSO₄ (magnesium sulfate)

Each student will use ~5 mL.

To make 1 L:

Dissolve 100 g MgSO₄ in 1 L tap water.

Quinine

Each student will use ~5 mL.

To make 1 L:

Quinine sulfate capsules can be purchased in a drug store. Dissolve the contents of 1 capsule (~0.45 g) in 1 L of tap water. Heating will improve the solubility, but you will need to strain the mixture through cheesecloth before use. Quinine sulfate can also be purchased from chemical supply houses.

10% NaCl (sodium chloride)

Each student will use ~5 mL.

To make 1 L:

Dissolve 100 g NaCl in 1 L tap water. Table salt can be used for this.

10% sucrose

Each student will use ~5 mL.

To make 1 L:

Dissolve 100 g sucrose in 1 L tap water. Granulated white sugar can be used for this.

Summary of Materials

	Per team of students	Per room or lab section
Disposables		
cotton swabs	3	
cups (3 oz.)	8	
gloves	2 pairs	
Life Savers	1 candy	
taste papers:		
control	1 square	
PTC	1 square	
sodium benzoate	1 square	
thiourea	1 square	
wax marker	1	
Equipment		
light sources (e.g., slide projector)		2
water bath (optional)		1
Labware		
1-L beakers	3	
100–1000-mL bottles (screw cap)		6
compass	1	
dissecting pan	1	
filters (red, green)		1 each
ice bucket		1
probe (blunt)	1	
30-cm (12") ruler	1	
thermometer	1	
white paper		1 sheet
Preserved Specimen		
sheep eye	1	
Solutions/Chemicals		
5% acetic acid	~ 5 mL	
10% $MgSO_4$	~ 5 mL	
quinine	~ 5 mL	
10% NaCl	~ 5 mL	
10% sucrose	~ 5 mL	
Other		
color blindness test booklet (optional)		1

LAB TOPIC 17

Plant Structure

✳ NOTE: Clover or radish seeds must be germinated 3 days before they are needed for the lab topic. See Exercise 17.3, Activity B, for details.

Students will work either alone or in pairs for this lab topic. Check with the instructor. Set up a workstation for each student or pair of students.

Each student or pair of students will need a compound microscope to use throughout the lab session. Use of a dissecting scope is suggested for a couple of procedures, but they could be shared among several students or pairs.

In addition to the materials listed for each exercise, each lab room should have:

- A waste stain container for disposal of toluidine blue
- A container for disposal of used razor blades
- A container for disposal of coverslips

Exercise 17.1: Plant Cells and Tissue Systems

No materials are necessary for this exercise.

Exercise 17.2: Plant Organs

Activity A: Leaves

Materials for each student or pair

- Prepared slide of dicot leaf cross section: The text in the Lab Investigations manual describes a mesophytic dicot leaf cross section. *Ligustrum* ("typical dicot leaf") from Carolina Biological Supply (#B 598D) is an excellent slide; it is well stained and it clearly shows distinct palisade and spongy parenchyma. This slide is also used for Exercises 17.3, 17.4, and 17.5. Many other species (for example, lilac and privet) can also be used.

Activity B: Roots

Materials for each student or pair

- Prepared slide of dicot root cross section: The text in the Lab Investigations manual describes a dicot root cross section from an annual plant. Slides showing mature *Ranunculus* (buttercup) root (e.g., Carolina Biological Supply #B 520) are typically used. This slide is also used for Exercises 17.3, 17.4, and 17.5.

Activity C: Stems

Setting up paraffin blocks at the beginning of the lab session

Materials for the instructor

■ Nuts and bolts: 1 for each student, pair, or team in a lab section plus several extra. Nuts that are 0.5" deep and have a 0.5" internal diameter work well. The nuts and bolts can be obtained at any hardware store. Be sure to get flat-headed bolts so the assembly can stand alone on the table. After all the lab sessions are over, clean the paraffin out of the nuts and bolts by soaking them in a solvent. Xylene is typically used as a solvent for wax. Fisher Scientific sells a product called Wax Out (#SW20-4) that is nonflammable and nontoxic.

 SAFETY NOTE: If you use xylene, carefully read the Materials Safety Data Sheet from the manufacturer before use. Xylene is flammable and should not be inhaled. Use only in a fume hood and wear gloves.

■ 100-mL beaker: Will be used to melt paraffin. Since it is difficult to get the paraffin cleaned out of the beaker, set aside paraffin beakers that can be used year after year.
■ 250-mL beaker with the water bath for melting the paraffin

 SAFETY NOTE: Never set a beaker of paraffin directly on the hot plate. It is flammable.

■ Hot plate
■ Block of paraffin: Can be found in the grocery store with home canning supplies. Histological paraffin (e.g., Histowax from Carolina Biological Supply, #86-7330) is more expensive but is superior for this exercise. Histological waxes are supplied in flake or pellet form and have a lower melting point.
■ Cutting board
■ Sharp knife (for slicing paraffin off the block; not needed if histological paraffin is used)
■ *Coleus* plants (0.5" stem for each block): Any type of *Coleus* can be used, and either the primary stem or lateral branches are suitable. Local greenhouses stock *Coleus* bedding plants (~ 3–4" tall) in the spring. At other times of the year, you may use older plants that have been kept in a greenhouse. Since each student or pair requires only 0.5" of stem, a large plant can supply numerous lab sections with tissue. Although the Lab Investigations manual text describes *Coleus* stems in some detail, sunflower (*Helianthus*) or bean (*Phaseolus vulgaris*) seedlings can be substituted if necessary.

Preparing the Tissue Blocks

These instructions are also in the Annotated Instructor's Edition. Check with the course instructor about whether the prep staff or the instructor will be responsible for performing this procedure. It takes about 30 minutes to prepare the paraffin blocks. If you are not doing it immediately before the lab session begins, be careful to seal the top of the tissue with paraffin or it will dry out.

1. Put small chunks of paraffin in a beaker. If you are using histological paraffin, it is already in pellet form. Put that beaker inside a larger beaker containing water. Place this "double boiler" on a hot plate to melt the paraffin.

> ⚠️ CAUTION: Paraffin is flammable. Do not set the beaker containing paraffin directly on the hot plate.

2. Barely thread the nuts onto the bolts to leave the maximum depth inside.
3. Cut pieces of stem tissue approximately 0.5" long, making the cuts as perpendicular to the stem axis as possible. Any straight section of the stem will do, including the nodes.
4. Set each stem piece upright in the center of a nut.
5. Pour melted paraffin into each nut. The stem section will stay upright. The paraffin must cover the top of the tissue or it will dry out.
6. The paraffin hardens in about 15 minutes. Do *not* attempt to hasten the process with ice or cold water or the paraffin may crack.

Materials for each student or pair of students to make stem sections

You can either give each team dropping bottles of ethanol and toluidine blue stain or have larger stock bottles in a central location and let students come and fill their dishes from those. If you use stock bottles, tape a test tube to the side of each bottle and put a disposable pipet in it for students to use to obtain their solutions. They don't need to measure the solutions. Students will need dropping bottles of water for other exercises.

- Bottle of distilled water, 5 mL per student
- Bottle of 50% ethanol, 5 mL per student: 1:1 (v/v) solution made with tap water.
- Bottle of toluidine blue stain, 5 mL per student: See the Recipe section. Provide a container for collecting waste stain.
- 50% glycerine, 1 mL per student in dropping bottle: 1:1 (v/v) solution made with tap water.
- 3 small Petri dishes: Small (60 × 15 mm) Petri dish tops or bottoms (Carolina Biological Supply #74-1346) are convenient, but any similar-size dish can be substituted (for example, syracuse watch glass, Carolina Biological Supply #74-2320). Small dishes are preferable because students will use less stain than they would if they were given larger dishes.
- Single-edged razor blade: Use brand-new single-edged razor blades. Carolina Biological Supply sells them (#62-6930); you can buy them slightly cheaper in a hardware store. Be sure to provide a container for collecting used blades.
- Forceps
- 4 microscope slides and coverslips

> ✳ NOTE: If the instructor wants to use a prepared slide of stem sections, be sure to use stems of annual, not perennial, plants. Carolina Biological Supply sells *Coleus* cross sections (#B 562) as well as *Helianthus* (#B 558) and *Ranunculus* (#B 576).

Exercise 17.3: Dermal Tissue

Activity A: Dermal Tissue in Stems

No additional materials are needed for this activity.

Activity B: Dermal Tissue in Roots
Materials for each student or pair of students

- 1 germinated clover, rye, or radish seed: Clover, radish, or any other small, rapidly germinating seeds can be used for this exercise. To germinate the seeds, place them on several layers of wet filter paper in a Petri dish, cover the Petri dish, and put it in the dark for 3 days. Check periodically and keep the filter paper moist but not soaking wet.
- Microscope slide and coverslip

Activity C: Dermal Tissue in Leaves
Materials for each student or pair of students

- ~½ leaf (depending on size): *Zebrina* (zebra plant) and *Rhoeo* (oyster plant) leaves are often used for epidermal peels because the epidermis is easy to peel and the epidermal cells on the underside of the leaf contain anthocyanins, which make them easy to see. Any leaves that are easy to peel are suitable. Unfortunately, it is difficult to obtain peels from *Coleus* leaves.
- Forceps (curved forceps are preferable)
- Water in dropping bottle
- Razor blade (can use the same one that was used for Exercise 17.2)
- Slide and coverslip

Exercise 17.4: Ground Tissue

Activity A: Parenchyma

No additional materials are needed for this activity.

Activity B: Collenchyma

No additional materials are needed for this activity.

Activity C: Sclerenchyma
Materials for each student or pair of students

- Pear: Each student will use only a tiny amount of the pear flesh. Any type of fresh pear can be used.
- Teasing needle
- Water in dropping bottle
- Slide and coverslip

Exercise 17.5: Vascular Tissue

No additional materials are needed for any of the activities in this exercise.

Recipe

Toluidine Blue Stain

Each student or pair of students will use ~5 mL.

Toluidine blue stain is made up in a $M/15$ phosphate buffer (pH 6.8), which is made as follows:

Solution A

9.47 g Na_2HPO_4 (anhydrous sodium phosphate dibasic) brought up to 1 L with distilled water.

Solution B

9.08 g KH_2PO_4 (anhydrous potassium phosphate monobasic) brought up to 1 L with distilled water.

For every 1000 mL buffer needed, mix 495 mL Solution A with 505 mL Solution B.

To make the stain, dissolve 0.5 g toluidine blue O in 1000 mL phosphate buffer. Dilute 1:1 with distilled water before use. The solution should be good for at least a year.

Provide a waste container in the lab room for excess stain.

The toluidine blue recipe, which is the one we use, is from Rock, B. "Three-dimensional plant anatomy via hand sectioning and differential staining." In J. Glase (ed.), *Tested Studies for Laboratory Teaching*, Vol. II. Dubuque: Kendall/Hunt, 1981, pp. 1–15.

Other sources, which are listed in the Annotated Instructor's Edition, suggest using a 0.1M or 0.01M phosphate buffer at pH 6.8 to make up the stain or simply making a 0.05% solution of toluidine blue O in distilled water.

Summary of Materials

	Per student or pair of students	Per room or lab section
Disposables		
Coleus stem	0.5" piece	
germinated seed (e.g., clover or radish)	1	
leaf (e.g., zebra plant, oyster plant)	½	
paraffin		~1 oz.
pear		small piece
single-edged razor blade	1	
Equipment		
compound microscope	1	
dissecting microscope (optional)	1	
hot plate		1
Labware		
100-mL beaker		1
250-mL beaker		1
cutting board		1
dropping bottles	4	
forceps	1	
sharp knife		1
slides and coverslips	7	
small Petri dish halves	3	
teasing needle	1	
used razor blade container		1
used coverslip container		1
waste stain container		1
Prepared Slides		
Leaf cross section	1	
Root cross section	1	
Solutions/Chemicals		
distilled water	~10 mL	
50% ethanol	~5 mL	
toluidine blue stain	~5 mL	
50% glycerine	<1 mL	

LAB TOPIC 18

Flowers, Fruits, and Seeds

✳ NOTE: Corn and bean seedlings must be planted 4–7 days before they are needed for this lab topic. See Exercise 18.3, Activity B, for details.

Exercise 18.1: Floral Structure

Activity A: Flower Dissection

Materials for each student team

- Flower: Any kind of flower that is available in abundance and has all four parts (petals, sepals, stamens, and carpels) can be used for the dissection. When we do this lab topic in the spring, fruit trees (crabapples, pears, and so forth) meet these criteria. At other times we purchase gladioli from a greenhouse. Geraniums work well. Check with the course intructor regarding available flowers. There are certain types you should avoid because it is difficult for students to interpret their parts (for example, anything related to daisies).
- Razor blade
- Teasing needle
- Dissecting microscope (optional)

Activity B: Pollination

Materials for the lab room

- Assortment of flowers: Ideally, you should collect fresh flowers for this exercise (if you can't, skip down to the next paragraphs). The purpose of the exercise is to have students examine a variety of flowers, so get as many different types as possible, including the flowers of trees and grasses. You don't need to look for the big showy varieties from gardens; get the little weeds that grow in lawns and sidewalk cracks. Consult with the instructor about the number of different types he or she would like to have on display. Each team of students will only have time to examine about six or so during the lab session. Most flowers will last quite well for a week of lab sessions if you put them in a small flask of water. If you know the name of the flower, put a label on it. You may want to put signs on the flasks indicating whether or not students may dissect the flowers.
- Borrowed flowers: If you can't collect flowers, you may be able to borrow plants from your school's greenhouse. The exercise asks students to hypothesize about how each flower is pollinated based on its color and odor, so try to get several different types.
- Slide set or pictures: If no flowers are available, the instructor may want to use a set of 35mm slides for this exercise. Carolina Biological Supply # 48-2874 was compiled to show different pollinators, but any set of flower slides can be used.

Figure 18.1
Preparing a rubber stopper with a slit
at one end.

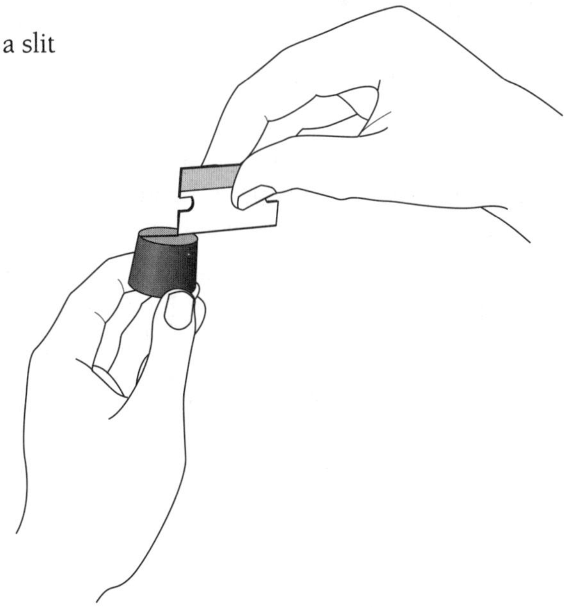

- Dissecting tools: Razor blade, teasing needle, and dissecting scope from Activity A (needed only if students are permitted to dissect the flowers).
- #0, 1, or 2 rubber stopper with slit at one end: This will be used to hold a flower upright for dissection (see Figure 18.1). Use a razor blade to make a slit about $1/4$" deep all the way across the small end of the stopper.

Activity C: Pollen Germination

Materials for each student team

If flowers are in short supply, the instructor may do this activity as a demonstration.

- Ripe stamen: In this exercise students will attempt to germinate pollen on a microscope slide. Two types of flowers that work well for this are *Vinca* (periwinkles) and *Impatiens*; both are commonly available in greenhouses. If the stamens are ripe, the anthers are open and you should see pollen grains coming out. Each team will need one stamen, so you should supply several flowers per lab section. The pollen of many plants will germinate *in vitro*, but it may be difficult to find a species that will germinate during the lab period. You may want to test plants that are more readily available to you than the species mentioned. A useful list can be found in Schimpf, D. J. "Rapid germination of pollen *in vitro*." *American Biology Teacher*, 1992, 54:168–169.
- Forceps
- Sucrose solution in dropping bottle: Different species of flowers require different concentrations. *Vinca* germinates in 20% sucrose. *Impatiens* germinates in 10% sucrose, which is typical of many species (although members of the Composite family require 30%–40% sucrose).
- Concavity (depression) slide (e.g., Carolina Biological Supply #63-2200)
- Compound microscope

Exercise 18.2: Fruits and Seeds

Activity A: Classifying Simple Fleshy Fruits
Materials for the lab room

In this exercise, students will use a key to identify botanical types of fruit. Following is a list of suggestions for each fruit type. Check with the instructor about how many examples to use for this display. The instructor may also suggest additional locally available varieties.

All fruits should be cut in half transversely (crosswise) and wrapped in cellophane or placed in plastic bags to retard spoilage. You can save fruit by putting only one half in a display, except for the drupes (students must see the seed). A single display should be sufficient for the lab room, but the instructor may want to have it split into different locations to avoid crowding. The fruits should *not* be labeled as to type; students are supposed to figure out whether a fruit is a "berry" or a "drupe."

Berry
> tomato
> eggplant
> all peppers
> grape
> chili pepper

Hesperidium (any citrus fruit is in this category)
> orange
> lemon
> lime
> tangerine
> grapefruit

Pepo (any melon or squash is in this category)
> squash
> cucumber
> pickle
> watermelon (1 slice is enough)
> okra
> cantelope
> pumpkin

Drupe
> avocado
> olive
> cherry
> plum
> apricot
> date

Pome

apple

pear

Strawberries, blackberries, raspberries, pineapples, and figs are not "simple" fruits; thus they cannot be identified using the student key. In addition, bananas will not "key out" correctly and so should not be used.

Activity B: Seed Dispersal

Materials for the lab room

Create a display of seeds and fruits that illustrate dispersal mechanisms. A seed dispersal set is available from Carolina Biological Supply (#15-7970). You can make up your own display or supplement the one from Carolina Biological Supply by collecting seeds as they become available over the course of a year. The idea is to collect seeds and fruits representing different dispersal mechanisms. You don't necessarily have to go to the country to collect; many of these can be found on campus or in vacant city lots.

- Wind-borne seeds: Dandelions (and related species with a similar fluffy head), milkweeds, maple, ash, elm, oleander, wild lettuce, thistles and oyster plants are good examples of wind-borne seeds.
- Seeds that stick to animals: Seeds that are dispersed by sticking to the fur of animals include cockleburs (burdocks), sand burs, stick-tights, beggar ticks, devil's claw, and hound's tongue.
- Fruits that are eaten by animals or are distributed when animals bury them: These include nuts (hickory nuts, hazelnuts, acorns), wild cherries, mulberries, quinces, crabapples, holly berries, ground cherries—just about any fruit that animals eat. You can buy fruits such as cherries, blackberries, apples, and plums.
- Water-borne seeds: There are a few fruits that are dispersed by water, including coconut and the American lotus, which is quite common in ponds and lakes.
- Other dispersal methods: You may also be able to find a few seeds that are mechanically dispersed such as *Impatiens* (touch-me-not) and wood sorrel. These don't work too well on the display, however, since once the seeds are released you can't put them back.

Put the seeds in Petri dishes, glass jars, or plastic bags for the lab session. If you can, identify the plant the seeds came from and label it for the students, but it isn't necessary to know the name in order to include it in the display.

The dry fruits and seeds mentioned above will keep indefinitely if stored away from moisture in zipper-type plastic bags or jars. To build your collection, collect whenever you have the chance and gather large numbers of seeds. Expect that handling by students will diminish your stocks each year. The Annotated Instructor's Edition suggests that students can be asked to bring in seeds from a scavenger hunt; this can help you build and maintain your collection.

Exercise 18.3: Seeds and Seedling Development

Activity A: Seed Structure

Each student will need at least 1 peanut (roasted in the shell) for this exercise. Buy plenty of peanuts, since students will eat them. Provide trash containers for every bench.

Activity B: Seedling Development

Materials for each student team

- Bean seed: Soak the beans overnight in the refrigerator to soften the seed coats. Use a large type of bean such as lima or kidney beans.
- Corn seed: Soak the corn overnight to soften the seed coat.

⚠ CAUTION: Don't use corn or beans that have been coated with a fungicide. Students will be handling the seeds.

- Razor blade
- Clear plastic cup containing soil, corn seedling, and bean seedling: The seedlings should be planted 4–7 days before they are needed for the lab session. To plant them, first soak the seeds in water overnight. Fill a *clear* plastic cup (the short 6-oz. size works well) with dampened soil. Push a bean down about ¹/₂" below the surface on one side of the cup and a corn seed on the other side. Both seeds should be visible through the side of the cup (see Figure 18.2). Plant enough cups so that each team in a lab section will have one (plus some extras in case seeds don't germinate). The same cups can be used in successive lab sections.

Figure 18.2

How to plant the bean and corn seedlings. (Seeds are shown side by side only for the purposes of the illustration; they may be planted on opposite sides of the cup.)

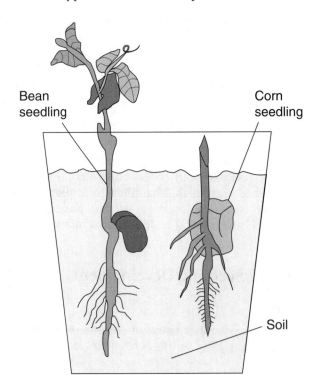

Summary of Materials

	Per team of students	Per room or lab section
Equipment		
compound microscope	1	
dissecting microscope	1	
Labware		
concavity slide	1	
cup (6-oz. clear plastic)	1	
forceps	1	
razor blade	1	
rubber stopper with slit (#0, 1, or 2)	1	
teasing needle	1	
Fresh Materials		
bean seeds	2	
corn seeds	2	
flower for dissection	1	
flower display (see discussion on p. 94)		
fruits		5–10
peanuts	2	
seed dispersal display (see discussion on p. 97)		
soil-filled plastic cup	1	
stamen from flower	1	
Solutions/Chemicals		
10–20% sucrose solution	~1 mL	

LAB TOPIC 19

Animal Behavior

In this lab topic students perform two demonstration exercises using pillbugs (Exercise 19.1) and bettas (Exercise 19.2). In Exercises 19.3 and 19.4 students design and perform their own experiments based on what they have learned from the demonstration exercises.

Since this lab topic includes the use of living fish, be sure that you are informed on your institution's animal use policy.

Exercise 19.1: Kinesis in Pillbugs

Materials for each student team

- Pillbugs in terrarium: Pillbugs can be ordered from Carolina Biological Supply (#L624, L630A, or L630D can be used; it doesn't matter whether you get pillbugs or sowbugs). Pillbugs can also be field-collected. They are found in dark, moist places such as the undersides of rocks and logs. Prepare a terrarium to keep them in. Pillbugs require high humidity, and they like to have rocks or pieces of wood to crawl under. If you don't want to set up an actual terrarium, you can use moist, crumpled paper towels in a glass or plastic container. Be sure to put a lid on the container or the pillbugs will crawl out. Each team of students will use approximately 10–20 pillbugs during the lab period. After they have been used in experiments, the pillbugs should be placed in a recovery terrarium. They can be reused after a rest. Let them rest through the next lab period (or overnight) before reusing them.
- Recovery terrarium for pillbugs: Have a second container prepared for pillbugs to rest between uses. It should also provide moisture and cover for the pillbugs and should have a lid. Label it with a sign.
- Choice chamber: Figure 19.1 shows how to construct a choice chamber. It consists of two large disposable Petri dishes (either 100-mm or 150-mm diameter) with a piece removed from the side of each one. To remove the piece, heat a knife or screwdriver in a flame. Apply the heated knife to the edge of the dish and press gently to make a slit in the plastic down to the bottom. Make a second slit about an inch away from the first. Then snap off the piece in between the two slits. Place two Petri dishes so that these doorways are aligned, and attach them by means of cellophane tape applied to the bottom. Use clear, not colored or opaque, tape.
- Small brush: An artist's brush or similar (e.g., Carolina Biological Supply #70-6192), used by students for transferring pillbugs from one container to another.
- Regular size Petri dish with filter paper disk in it.
- Watch or clock with second hand: There will probably be enough students who have watches, but it is convenient to have a wall clock with a second hand located where all students can readily see it.

Figure 19.1
Construction of choice chamber.

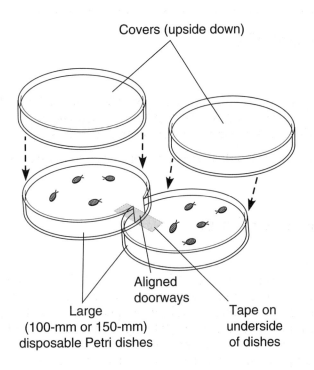

Covers (upside down)

Aligned
doorways

Large
(100-mm or 150-mm)
disposable Petri dishes

Tape on
underside
of dishes

Exercise 19.2: Agonistic Behavior in Siamese Fighting Fish

Materials for each student team

- Male Siamese fighting fish (*Betta splendens*) in individual containers: Each team needs 1 fish. Siamese fighting fish, often referred to as bettas, are sold in pet stores. Carolina Biological Supply also sells them, but at more than twice the typical pet store price. Choose brightly colored fish; a pale fish may not be healthy. All the fish must be males. These fish will injure each other if placed in the same tank, so you need to have a separate container for each fish. (Bettas don't do well in the same aquarium with other species, either, as other fish tend to bite the betta's fins and damage them.) Put a number in an unobtrusive place on each container. Small, flat-sided fish tanks (e.g., Carolina Biological Supply #67-0430) are preferred, because the flat sides work best for the experiments students will do. However, if you don't want to buy and store fish tanks, 1-L or 2-L beakers can also be used. To prepare the tank, put some aquarium gravel on the bottom, and add a few snails if you're planning to keep the fish for a while. You can either condition tap water by letting it stand for several days or buy bottled *spring* water. (Don't use distilled water or "drinking" water.) You'll also need tropical fish food to feed the bettas. Aeration is not necessary since bettas come to the surface to get air. You may be able to find a pet store that will sell you bettas and then buy back the fish that remain after the lab topic (which should be most of them). If it is more convenient for you to buy new fish each term than to maintain them, consider letting your students adopt the bettas after the lab topic is completed or, if your institution's accounting procedures permit it, buy them from you.
- Small mirror: 3" × 4" or 4" × 4" size, or large enough for a betta to see its reflection. A good source for mirrors is the cosmetics section of a discount store.

Exercise 19.3: Designing an Experiment

No materials are needed for this exercise.

Exercise 19.4: Performing the Experiment and Interpreting the Results

The basic kit for student-designed investigations consists of the materials used in Exercises 19.1 and 19.2. Instructors may allow students to choose whether they want to do experiments with pillbugs or bettas, or else they may assign one animal or the other to each team. If half of the teams work with each animal, the number of pillbugs and bettas available from Exercises 19.1 and 19.2 will probably be sufficient. If students choose their own subject animal, additional animals will probably be needed. Check with the instructor about how many extra animals will be needed for the student-designed investigations. Also, review the following list of suggested additional materials with the instructor.

Materials for pillbug experiments

- Additional filter paper: Circles 5.5 cm, 7.5 cm, or 9 cm can be used.
- Black cloth or sheet of black paper (to darken one side of the choice chamber)
- Colored paper to change background color: If your lab benches are black, white paper should be included. If the benches are white, include black paper. Any selection of assorted other colors can be used. Sheets of paper should be at least as large as the area of the choice chamber.
- Indicating Drierite (Carolina Biological Supply #85-8950): A teaspoon of Drierite can be used to lower the humidity in one chamber. Use indicating Drierite, which changes color from blue to pinkish purple when it has absorbed water. Supply *blue* Drierite to the lab sections. Don't discard the Drierite when it's pink, however. Pink Drierite can be returned to its dry (blue) state by placing it in an oven at low heat to drive off the moisture.
- Insect repellant (liquid form)
- Small Petri dish top or bottom (to hold Drierite)

Materials for betta experiments

- Colored paper: Each team will probably use no more than a 3" or 4" square of each color. The colors should include red, blue, yellow, and green, but could also include other colors. Try to find brightly colored paper; some construction paper is rather dull. You'll save paper if you precut sheets of 8½" × 11" paper into 6 equal pieces so that students will use only as much as they need. Aluminum foil can also be included. The instructor may specify some other colors as well.
- Scissors
- White thread (1 spool can be shared by the lab section)
- Cellophane tape (1 roll can be shared by the lab section)

Summary of Materials

	Per team of students	Per room or lab section
Disposables		
spring water	~1 L per fish	
Labware		
brush (small)	1	
clock		1
fish tank or 1-L or 2-L beaker	1	
mirror	1	
Petri dishes (large)	2	
Petri dish (regular)	1	
terraria (for pillbugs)		2
Living Specimens		
pillbugs (in terrarium)	10–20	
betta (in tanks or beakers)	1	

LAB TOPIC 20

Foraging Strategy

For this lab topic, students will work in teams of 4 or 5. After completing Exercises 20.1, 20.2, and 20.3, each team will design and perform its own investigation. A few additional materials that students might require are listed below. Consult with the course instructor regarding any other materials that he or she would like to have available.

Except for the nail polish, all the materials can be used by all of the lab sections, though you may have to replace beans that are lost or broken.

Exercise 20.1: The Simulation System

No materials are required for this exercise.

Exercise 20.2: Single Prey Species and Exercise 20.3: Mixed Prey Species

Materials for each student team

- 2 plastic dishpans, ~12" × 14" × 6": The exact size doesn't matter. One dishpan should have sand in it, the other is empty.
- Sand, ~10 lb per dishpan: Remember, only 1 pan for each team has sand in it. Ask your hardware or building supply store for the type of sand that has the least dust. Sand that is sold for filling children's sandboxes has usually been washed to remove dust; it is a good choice for this lab topic.
- Bench protection: Get something to spread under the dishpans to catch any sand that spills so it doesn't scratch the bench tops. Large sheets of newsprint have the advantage that the spilled sand can easily be poured back into the dishpan. Terrycloth towels are good because they keep the sand from spilling from the benchtop onto the floor, but they must be taken outside and shaken.
- 200 dried kidney beans in a plastic bag (exact count not necessary)
- 50 dried lentils in a plastic bag (exact count not necessary)
- 50 dried Great Northern beans in a plastic bag (navy beans can be substituted; exact count not necessary)
- 5–10 dried lima beans in a plastic bag
- Watch or clock with second hand: Students can use their own watches. Only 1 is needed for each group. A wall clock that has a second hand can be used instead. Supply stopwatches if you have them, but they aren't necessary.
- Dish to collect beans: A 4.5" diameter culture dish works well. Don't use Petri dishes, as they are too shallow.
- Soil sieve or colander (can be shared among lab teams): After each investigation, students will pour the sand through a sieve or colander into the empty dishpan to remove any remaining

beans. The mesh on the sieve must be large enough to allow the sand to flow through easily, but small enough to retain the lentils.

- Brush and dustpan (for students to sweep up spilled sand at end of lab period)

Exercise 20.4: Designing an Experiment and Exercise 20.5: Performing the Experiment and Interpreting the Results

The basic kit for this investigative lab topic consists of the materials that students used for Exercises 20.2 and 20.3.

Materials for possible student investigations

- Fingernail polish: red and clear
- Pebbles: Smooth, bean-size pebbles. They can be purchased at pet stores, which sell them to use in aquariums. A bag that contains 100 or so pebbles should be enough for a lab section.
- More dried beans of the types used in Exercises 20.2 and 20.3: Have an extra bag of each on hand. The instructor may want to have other types of dried beans available as well.

Summary of Materials

	Per team of students	Per room or lab section
Disposables		
bench protection		to cover area
Fresh Materials (will be reused)		
Great Northern beans	50	
kidney beans	200	
lentils	50	
lima beans	5–10	
sand	10 lb	
Labware		
clock with second hand		1
dish	1	
dishpans	2	
sieve or colander	1 (can be shared)	
brush and dustpan	1 each (can be shared)	

Vendor Addresses

Barley & Vine
4 Corey Street
West Roxbury, MA 02132
1-800-666-7026 or 1-617-327-0089

Carolina Biological Supply Company
2700 York Road
Burlington, NC 27215
1-800-334-5551
Fax: 1-800-222-7112

Reference numbers are from the 1994 catalog.

The Cellar
Dept. ZR
P.O. Box 33525
Seattle, WA 98133
1-800-342-1871 or 1-206-365-7660

Edmund Scientific Company
101 East Gloucester Pike
Barrington, NJ 08007-1380
Order by phone: 1-609-573-6250
Order by fax: 1-609-573-6295
School orders/educational pricing: 1-609-573-6270

ESHA Research (Lab Topic 14, Alternative Exercises)
Nutrition Analysis Software
P. O. Box 13028
Salem, OR 97309
Fax: 1-503-585-5543
Phone: 1-503-585-6242

Fisher-EMD (Educational Materials Division)
4901 West LeMoyne Street
Chicago, IL 60651
1-800-621-4769
Fax: 1-312-378-7770

Reference numbers are from the 1993/94 catalog.

Fisher Scientific
711 Forbes Avenue
Pittsburgh, PA 15219-9919

Catalog lists toll-free ordering numbers by state
Reference numbers are from the 1993/94 catalog.

The Home Brewery
    (locations in Missouri, California, and Nevada)
    1-800-321-BREW (2739) for catalog

Intellimation Library for the Macintosh (Lab Topic 14, Alternative Exercises)
    Department YAS
    P. O. Box 1922
    Santa Barbara, CA 93116-1922
    1-800-346-8355
    Fax: 1-805-968-8899

Lactaid, Inc. (Lab Topic 14, Alternative Exercises)
    P. O. Box 111
    Pleasantville, NJ 08232
    Lactaid Hotline 1-800-257-8650 (9–4 weekdays, Eastern Standard Time)

Sargent-Welch Scientific Company
    911 Commerce Court
    Buffalo Grove, IL 60089-2362
    1-800-SARGENT
    Fax: 1-708-677-0624

    Reference numbers are from the 1993–95 catalog.

Sigma Chemical Company
    P. O. Box 14508
    St. Louis, MO 63178-9916
    1-800-325-3010 (ordering)
    1-800-325-8070 (customer service)
    1-800-325-5052 (Fax)

    Reference numbers are from the 1994 catalog.

Urban Technology
    2911 West Wilshire
    Oklahoma City, OK 73116
    1-800-356-3771 or 1-405-843-1888

Ward's Natural Science Establishment, Inc.
    P. O. Box 92912
    Rochester, NY 14692-9012
    1-800-962-2660
    Fax: 1-800-635-8439

    Reference numbers are from the 1994 catalog.

Planning Guide

This handy checklist will help you anticipate your needs for the entire course, organize your ordering, and plan your preparations. It contains each piece of equipment needed as well as all specimens, solutions, and chemicals, listed by their corresponding lab topic (including the Alternative Exercises, which appear in gray shading). Some materials, especially pieces of equipment such as Spec 20s, are optional; the Instructor Notes in the Lab Investigations manual give details. Materials that can be found in any laboratory, such as test tubes, beakers, pipets, and so forth, are not listed here.

PLANNING GUIDE

(shaded areas indicate Alternative Exercises; ▽ = optional)

Equipment	1	2	3	3A	4	5	6	6A	7	7A	8	8A	9	10	11	12	13	14	14A	15	15A	16	17	18	19	19A	20	20A
Alcohol lamp														•														
Balance		•					•																					
Blender							•	•	•	•																		
Blood pressure kit																				•	•							
Centrifuge (tabletop model)									•	•																		
Compound microscope						•								•	•	•					•		•	•				
Dissecting microscope						•								▽	•	•	•	•					▽	•				
Hot plate			•	•	•			•				•							•				•					
Ice chest																			•									
Light bank													•															
pH meter			•	•																								
Slide projector								▽	▽	▽												•						
Spectronic 20 or other spectrophotometer											•	•																
Stethoscope																		•		•	•							
Stir plate			▽	▽																								
Vacuum source											•	•										▽						
Water bath								•																				

Living or fresh specimens	1	2	3	3A	4	5	6	6A	7	7A	8	8A	9	10	11	12	13	14	14A	15	15A	16	17	18	19	19A	20	20A
Amphibian display																	•											
Angiosperm															•													
Angiosperm seeds															•													
Anther from flower															•									•				
Bean seeds																								•				
Chicken eggs (fertile)																	•											
Clam or mussel (fresh)																•												
Coleus																							•					
Corn seeds															•													
Daphnia or brine shrimp																•												
Dried beans																											•	
Earthworm																•												
Elodea						•																						
Fern															•													
Fern spores															•													
Fish eggs																	•											
Flower for dissection																								•				

Living/fresh specimens, cont.	1	2	3	3A	4	5	6	6A	7	7A	8	8A	9	10	11	12	13	14	14A	15	15A	16	17	18	19	19A	20	20A
Flower display																								•				
Frog eggs																	•											
Fruit display																								•				
Gingko															•													
Horsetail (Equisetum)						•									•													
Hydras																•												
Leaf																							•					
Liverwort															•													
Moss															•													
Moss spores															•													
Ovary from flower															•													
Paramecium						•																						
Peanuts																							•					
Pear																												
Pillbugs (sowbugs)																									•			
Pine branch															•													
Pine cones, female															•													
Pine cones, male															•													
Pine seeds															•													
Planaria																•												
Reptile display																	•											
Seed dispersal display																								•				
Seeds, germinated																							•					
Selaginella (club moss)																									•			
Siamese fighting fish (bettas)																												
Vinegar eel culture															•	•												
Zamia (cycad)																												

Solutions/Chemicals	1	2	3	3A	4	5	6	6A	7	7A	8	8A	9	10	11	12	13	14	14A	15	15A	16	17	18	19	19A	20	20A
Acetic acid, 5%																						•						
Acetic acid, 0.1M				•															•									
Acetic acid, 0.1N				•																								
Acetylsalicylic acid, 0.1M				•																								
Ascorbic acid, 0.1M				•																								
Barfoed reagent					•																							
Benedict reagent																			•									
Bile, 5%					•													•										
Biuret reagent								•																				
Buffer, pH 2			•	•																								
Buffer, pH 4			•	•																								

Solutions/Chemicals, cont.	1	2	3	3A	4	5	6	6A	7	7A	8	8A	9	10	11	12	13	14	14A	15	15A	16	17	18	19	19A	20	20A
Buffer, pH 6			•	•				•																				
Buffer, pH 7			•	•				•																				
Buffer, pH 8			•	•				•																				
Buffer, pH 10			•	•				•																				
Buffer, pH 12			•	•				•																				
Casein, 0.5%																			•									
Catechol, 0.5%							•	•																				
Chick Ringer's solution																												
Chloroplast pigments												•					•											
Chromatography solvent												•																
Citric acid, 0.1M				•								•																
Crystal violet stain														•														
DCPIP (dichlorophenol-indophenol), 1mM									•	•																		
Egg albumin, 1%					•																							
Ethanol, 50%																							•					
Ethanol, 70%												•								•	•							
Glucose, 0.1M					•																							
Glycerine, 50%																			•				•					
Gram's iodine														•														
Hydrochloric acid, 0.5%																		•	•									
Hydrochloric acid, 0.1N			•	•										•		•												
Hydrochloric acid, 2N					•																							
Iodine reagent					•							•						•	•									
Lactaid																			•									
Lactose, 0.1M																			•									
Magnesium sulfate, 10%																						•						
Methylene blue						•																						
Milk																			•									
Nutrient agar																												
Pancreatic extract														•				•	•									
Phenolphthalein, 1%																		•	•									
Phenol red			•	•					•	•				•														
Phosphate buffer, pH 7.2																												
Quinine																						•						
Safranin stain														•														
Sodium bicarbonate, 0.1M				•																								
Sodium bicarbonate, 1%																			•									
Sodium bicarbonate, 0.2%											•	•																
Sodium bicarbonate, 5%			•																									

Solutions/Chemicals, cont.	1	2	3	3A	4	5	6	6A	7	7A	8	8A	9	10	11	12	13	14	14A	15	15A	16	17	18	19	19A	20	20A
Sodium chloride, 10%																						•						
Sodium hydroxide, 0.1N			•	•										•				•	•									
Sodium phosphate, dibasic, 0.2M				•																								
Sodium phosphate, monobasic, 0.2M				•																								
Starch, 0.5%																		•	•									
Starch, 1%																												
Succinate, 0.025M					•				•	•																		
Sucrose, 10%																						•		•				
Sucrose, 0.1M					•																							
Sucrose-phosphate buffer									•	•																		
Toluidine blue O stain																							•					
Vegetable oil																		•	•									

Trademark Acknowledgments

- Adolph's® is a registered trademark of Adolph's Ltd.

- Alka-Seltzer is a registered trademark of Miles Inc.

- Arm & Hammer is a registered trademark of Church & Dwight Co. Inc.

- Ben-Gay is a registered trademark of Pfizer Inc.

- Big Mac is a registered trademark of McDonald's Corporation.

- Clorox® is a registered trademark of The Clorox Company.

- Dairy Ease, Lysol, and Milk of Magnesia are registered trademarks of Sterling Drug Inc.

- Drano is a registered trademark of The Drackett Company.

- Drierite is a registered trademark of W.A. Hammond Drierite Company.

- Fast Plants™ is a trademark of J.M. Marketing Services, Inc.

- Fleischmann's® is a registered trademark of Nabisco, Inc.

- The Food Processor is a trademark of ESHA Research Corporation.

- Fruit-Fresh and Tums are registered trademarks of Beecham Inc.

- Gatorade is a registered trademark of Stokely-Van Camp, Inc.

- Histowax® is a registered trademark of EM Industries Incorporated.

- Jell-O is a registered trademark of General Foods Corporation.

- Karo is a registered trademark of CPC International Inc.

- Kool-Aid is a registered trademark of Perkins Products Company.

- Krazy Glue® is a registered trademark of Krazy Glue, Inc.

- Lactaid is a registered trademark of Lactaid Inc.

- Life Savers is a registered trademark of Life Savers, Inc.

- Lime-A-Way is a registered trademark of Economics Laboratory, Inc.

- Maalox is a registered trademark of Rorer Pharmaceutical Corporation.

- Medeprin and Tylenol are trademarks of Johnson & Johnson.

- Motrin® is a registered trademark of The Upjohn Company.

- Nalgene® is a registered trademark of Nalge Company.

- NutraSweet is a registered trademark of NutraSweet Company, Inc.

- Parafilm is a registered trademark of Marathon Paper Mills Company.

- Pellon® is a registered trademark of Pellon Corporation.

- Pepto-Bismol is a registered trademark of Norwich Eaton Pharmaceuticals, Inc.

- Permount® (mounting glass for microscopes) is a registered trademark of Fisher Scientific Company.

- PineSol is a registered trademark of American Cyanamid Company.

- PROTOSLO® is a registered trademark of Carolina Biological Supply.

- Rolaids is a registered trademark of American Chicle Company.

- 7UP is a registered trademark of The Seven-Up Company.

- Spectronic 20 is a registered trademark of Bausch & Lomb Incorporated.

- Sprite® is a registered trademark of The Coca-Cola Company.

- Styrofoam is a registered trademark of The Dow Chemical Company.

- Sweet'n Low is a registered trademark of Cumberland Packing Corp.

- 10-K is a registered trademark of Suntory Water Group, Inc.

- Tygon® is a registered trademark of Norton Company.

- Wax Out™ is a trademark of Fisher Scientific Company.

- Whatman is a registered trademark of Whatman Paper Limited.